Introduction to
Petroleum Production

Volume 3

Gulf Publishing Company ● Book Division ● Houston, London, Paris, Tokyo

Introduction to Petroleum Production

Volume 3

Well Site Facilities: Water Handling, Storage, Instrumentation, and Control Systems.

D. R. Skinner

Introduction to Petroleum Production
Volume 3

Well Site Facilities: Water Handling, Storage, Instrumentation, and Control Systems

ISBN 0-87201-769-9

Volume 1
Reservoir Engineering, Drilling, Well Completions
Volume 2
Fluid Flow, Artificial Lift, Gathering Systems, and Processing
Volume 3
Well Site Facilities: Water Handling, Storage, Instrumentation, and Control Systems

Contents

v

Electronic Instrumentation. Chromatographs. Atmospheric Emissions Instruments. Vibration Measurement. Load Measurement. Annular Fluid Level Measurement. Electric Power in Production Operations.

Chapter 1
Centralized Treating

Volume 1 of this series covered the nature of oil and gas, how they are created and stored in reservoirs, and how those reservoirs are found and exploited. It also briefly examined how fluids are brought to the surface. Volume 2 detailed the principles and technology involved in getting oil and gas out of the ground and ready for transportation to the refinery or gas processing plant.

Volume 3 describes treatment and injection technology and practices. Some of the same technology and practices are used in enhanced recovery methods. Electrical equipment and systems and instrumentation and control equipment are also covered.

The Battery System

On some large leases of more than 20-30 wells, it is economically attractive to use small, independent treating facilities to serve a few wells. However, on most large leases it is much better to use large, centralized treating facilities to minimize capital expenditures on high-cost items such as treaters, tanks, and high-volume pumps.

When many wells are to be served by a process facility, the gathering system may become complex and require large-diameter pipe or complicated header designs. Also, single processing facilities can cause lease production losses when maintenance or repair work is done and wells and flowlines are blocked to stop flow. A combination of central and satellite batteries can be used to serve many wells and provide for maintenance and repair work without shutting down the entire lease. *Battery* is another name given to a treating and storage facility.

Satellite Batteries

Satellite batteries are small process facilities which accumulate, initially process, and temporarily store fluids produced from a segment of a large lease. Satellite batteries are usually provided for groups of 10-30 wells, but larger and smaller batteries may be used. Flowlines from wells terminate in headers at the satellite battery. The number of satellite batteries on a lease and the number of wells served by each is determined by optimizing the cost of the batteries and the flowlines serving them.

Many satellite batteries have facilities to test individual wells, and all have equipment to separate gas and liquid so the two fluids can be metered. Other batteries perform some preliminary treating to ease the load on the central facility. Figure 1-1 is a diagram of some of the functions performed in a satellite battery.

Well Test Headers

Flowlines are routed to a satellite battery and into a header sometimes called a well test header. One type of test header is shown in Figure 1-2. This header design allows one well's flow to be routed into process equipment used to measure the production rates of fluids from one well, while the remaining fluid is processed through other equipment commonly called the production treating equipment. Measuring an individual well's production is called "testing" the well, and the equipment employed is the test equipment.

Figure 1-3 shows a header equipped with pneumatic operators which, in turn, are driven by solenoids (electrically powered gas valves). The solenoids are activated from an electric switch panel where a switch is turned on to test one well while allowing the other wells' production to flow through the production vessel.

Well Test Facilities

Wells can be tested by routing their production through a process vessel and measuring the flow of oil, gas, and water. To measure fluid flow rates, gas and liquid must be separated because no single meter is able to measure both. There are several vessels used to test wells.

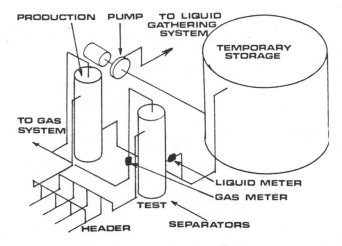

Figure 1-1. A satellite battery must allow for well testing, gas separation, temporary storage, and liquid transportation.

Figure 1-2. A well test header allows the produced fluids of one well to be routed to test equipment, while the fluids from other wells are processed by production vessels.

Figure 1-3. One type of automatic test header uses pneumatically operated valves to select wells.

Two-Phase Separators. The two-phase separator is one common test vessel. Gas is removed from the liquid and measured with one of the gas meters discussed in Volume 2 (see Chapter 2, Fluid Flow Measurement). The liquid is routed through a meter, and the total flow rate of oil and water is measured. A sample of the oil-water mixture is taken periodically to determine the *water cut,* the percentage of water in the liquid. This percentage is then applied to the liquid flow rate to determine oil and water flow rates. Sometimes a single sample is not an accurate enough record of water production, and an automatic sampler is needed to collect a few drops of liquid in a container for every barrel or tenth of a barrel of liquid that passes through the liquid meter. After the well test is complete, the contents of the sample container are analyzed for water cut. Figure 1-4 shows a two-phase separator with liquid and gas meters in service as a test vessel.

Treaters. Heater treaters are also used as test vessels. A vertical or horizontal conventional treater is installed with gas and liquid meters on the three fluid outlet lines as shown in Figure 1-5. Treaters are usually the most effective in separately measuring oil and water, but they are usually expensive. It is difficult to find a treater small enough to handle one well only.

Metering Treaters. Figure 1-6 shows the internal configuration of a metering treater. As oil and water separate as a result of heat, demulsifiers, and gravity, they collect in separate chambers. These chambers collect liquid in specific volumes (usually one barrel). When the float in a chamber switches, the diverting valve switches and allows liquid to dump. The float also activates a pneumatically operated counter. At the conclusion of a test, the two counters give an accurate indication of the volumes of oil and water produced.

Figure 1-4. A two-phase separator with meters on both liquid and gas outlets can be a test vessel.

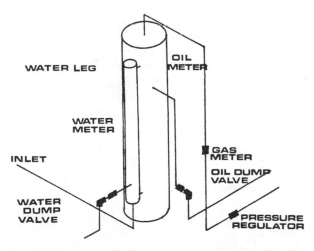

Figure 1-5. A vertical treater with three meters can be used as the test vessel.

Three-Phase Separators. Another method of measuring the oil and water production from a single well is to use a three-phase separator (Figure 1-7). This is simply a vertical, free-water knockout to which two liquid meters have been added. Incoming liquid strikes a splash plate and falls into the liquid section. An unweighted float periodically opens the oil dump valve, while a weighted interface float detects the top of the water layer and

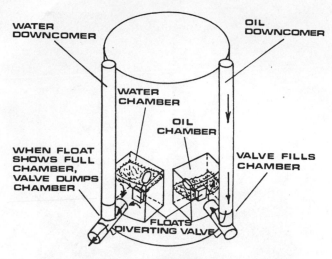

Figure 1-6. A metering treater has volume chambers which hold precise volumes and allow metering by recording the number of times the chambers are emptied.

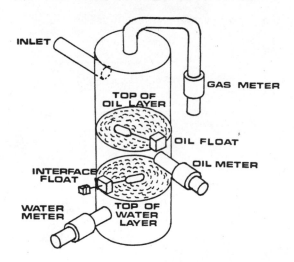

Figure 1-7. A three-phase separator with meters on all three fluid outlet lines makes a good test vessel.

operates the water dump valve. Some oil/water dispersions breakdown without difficulty, and demulsifiers may be added to make emulsions breakdown into oil and water. Three-phase separators are usually large to allow ample retention time for separation.

Portable Well Testers. Sometimes it is desirable to test an individual well at the wellhead. Figure 1-8 shows a portable well tester used for this purpose.

Figure 1-8. A portable tester can be a trailer-mounted test separator with liquid and gas meters. (Courtesy of Engelman-General, Inc.)

This small, trailer-mounted unit contains a two-phase separator, liquid and gas meters, and a sampler. If a wellhead is so configured, the tester is connected into the flowline with flexible hoses. Fluids are sent back into the flowline after being measured.

Production Vessels

In some satellite batteries the only function of the production facilities is to remove gas from the liquids. In this instance the only production vessel needed is a two-phase separator. In general, the fluids from a production vessel are not independently metered.

Sometimes it is advantageous to remove some (if not all) water from incoming liquid streams before they reach central facilities. Figure 1-9 shows a free-water knockout used to remove some water, but this vessel is not intended to treat oil for sales purposes.

The purpose of a satellite battery is usually to separate gas and store liquid. It is not often practical to attempt to separate oil and water for

Figure 1-9. A free-water knockout is sometimes used to separate some (if not all) water at a satellite battery.

anything more than testing. Generally, it is not possible to justify the cost of the equipment needed for efficient treating at small satellite batteries. However, on leases producing large volumes of free water, it is advantageous to separate some free water before transferring the liquid to a central battery.

Temporary Storage

Most satellite batteries are equipped with tanks to provide at least some liquid storage space. To optimize the cost of satellite batteries, tanks are selected which can store only a few hours' production from a lease. Again, in the concept of using satellite batteries as a part of a larger system, there is no need to provide for long-term storage, since storage can be handled at the central battery.

Fluid Transfer

Liquids and gases separated from each other at a satellite battery must be transferred to the central battery. In some satellites the pressure from the vessels is used to push liquid. This can be done when satellite batteries are located close to the central battery, but if the satellite batteries are far from

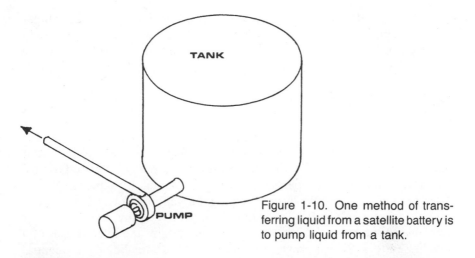

Figure 1-10. One method of transferring liquid from a satellite battery is to pump liquid from a tank.

Figure 1-11. Another method of transferring liquid is to pump it directly from a production separator.

the central battery, liquid must be pumped. Liquids from the test and production vessels can be dumped to a tank from which they can be moved with a centrifugal pump.

Liquid can be pumped from a tank (the conventional method shown in Figure 1-10) or directly from a separator (Figure 1-11). This method must be used carefully because the liquid in a separator still contains gas which can cause problems. However, if operated properly, this method is satisfactory.

Gas separated from liquid is usually satisfactory for processing in a gas processing plant or facility. There is no need to mix it with the liquids to send to the central battery. Gas is usually collected in a gathering system which is independent from the liquid system.

Satellite Pressure

Some satellites transfer liquid to a central battery with pressure from the vessels. For fluid to enter the vessels, the wellhead pressure must be higher than that of the vessels. The flowline pressure eventually is exerted against the reservoir. This is particularly important in high permeability, low-pressure reservoirs because increasing the wellbore pressure even a few psi can decrease fluid flow rate significantly. For many reservoirs, it is desirable to maintain the lowest possible pressure in the inlet vessels of a satellite battery. If this pressure is too low to transfer liquid to a central battery, a pump must be used. The cost of this pump may be justified by the additional hydrocarbon production caused by lower flowline pressure.

Satellite Equipment Arrangement

There are any number of arrangements of equipment in a satellite battery. Figure 1-12 shows a simple satellite battery in which the production and test vessels are both separators. One problem with this arrangement is that the liquid and gas meters on the production separator do not measure all the fluid from the battery. Figure 1-13 shows an alternate arrangement in which all production is measured.

A factor that is often overlooked when satellite batteries are designed is the effect of vessel pressure on the production rate of a well being tested. If a well is normally operated at the pressure of the production vessel but made to operate at a different pressure when routed through the test vessel, the result of the test will not be totally accurate since the flow rate changes as pressure changes.

When a well is switched from the production to the test vessel, the flowline pressure must increase to at least the vessel pressure. This pressure is exerted down the annulus against the reservoir. The increase in pressure can significantly reduce the fluid flow rate from the reservoir during the test, but the rate returns to normal when the well is switched back to the production vessel. The difference in pressure between the test and production vessels is the principal reason that the result of the tests for all wells is not the same as the normal satellite liquid flow rate.

In some satellites a duplication of equipment results if the vessels are arranged to maintain equal pressure in the test and production vessels. Figure 1-14 is a satellite battery in which test and production vessel pressures are equal, but no equipment duplication is required.

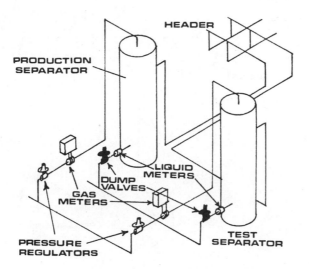

Figure 1-12. A test battery may consist simply of test and production separators and the associated meters.

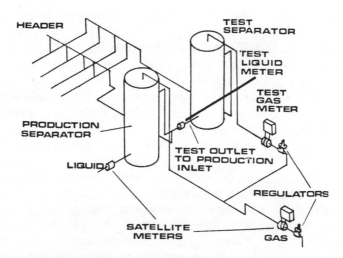

Figure 1-13. A satellite battery may be arranged so that all fluid is measured through the production meters.

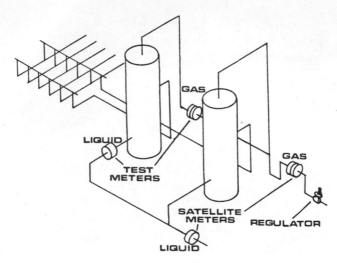

Figure 1-14. By using a single pressure regulator to control pressure in both vessels, the test and production separators operate at the same pressure.

Gathering Systems

Pipeline systems, called gathering systems, are used to collect fluids from satellites and transport them to central facilities. Liquids and gases are pumped into gathering systems at low pressures of 100 psi or less. The pressure required depends on the distance from the satellite to the central battery and the size of the pipe. Gathering systems may be constructed as radial systems (Figure 1-15) or trunkline systems (Figure 1-16).

Most gathering systems use large-diameter pipe to minimize friction losses. Because low pressures are utilized, inexpensive pipe of steel, plastic, fiberglass, or asbestos may be used. However, the fluids transported may be quite corrosive, and it is important that the pipe be resistant to corrosive attack. For this reason, fiberglass and plastic pipe are usually used.

On leases where large volumes of free water are produced, as much free water as possible may be removed at the satellite batteries. To avoid repeating the water separation process at the central battery, water is transported to the central battery in a gathering system separate from the oil, as shown in Figure 1-17. Using separate oil and water gathering systems requires two completely parallel systems, including pipe, pumps, tanks, and so forth. Water is seldom separated in this manner so that the cost of duplicate gathering and transportation systems can be avoided.

The gathering system is critical in the operation of centralized treating facilities. A leak in one of the gathering lines could require shutting down an

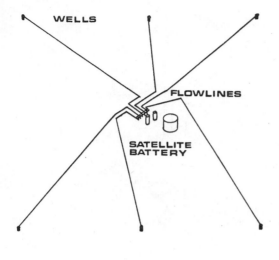

WELLS

FLOWLINES

SATELLITE
BATTERY

Figure 1-15. All flowlines are
routed to the satellite in a radial
gathering system.

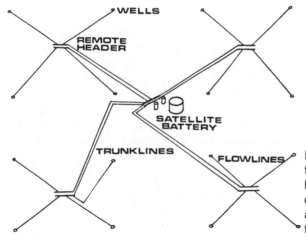

WELLS

REMOTE
HEADER

SATELLITE
BATTERY

TRUNKLINES

FLOWLINES

Figure 1-16. Flowlines
terminate at remote
headers and trunklines
carry fluids to satellites in
a trunkline gathering system.

entire lease if some method is not provided to isolate the line. Also, since the gathering lines are large, a significant volume of liquid can be lost from a single line if a leak develops. Radial gathering systems are usually terminated in headers equipped with valves that allow isolation of a single line while allowing others to function normally. Gathering systems may also be segmented by installing intermediate valves. In the event of a leak these valves may be used to isolate a section of a line to prevent the loss of the entire volume of the line. Trunkline gathering systems are usually segmented in the same way to allow isolation.

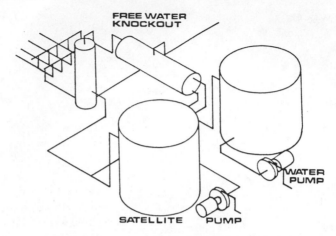

Figure 1-17. Some satellite systems separate water from oil and transport these two liquids separately.

When earth-working equipment is operated on a petroleum producing lease, there is always the possibility of striking and damaging pipelines. It is not desirable to damage any fluid-carrying line, but gathering lines are usually considered more critical than others. Because of their light, non-metallic construction, fiberglass, plastic, or asbestos gathering system lines are more difficult to locate than steel pipelines. Thus, gathering lines are usually buried deeply. The routes of these pipelines should be clearly and frequently marked with signs stating the type, depth, and contents of the pipe.

Central Batteries

Most fields with centralized process facilities have a single central battery sized large enough to process all fluid produced from a lease. Some of these facilities are intentionally designed to handle more fluid than is actually processed to avoid overtaxing any part of the system. More often, however, central batteries are designed for actual production, but provisions are made for expansion. Some leases are so large (several hundred wells and thousands of barrels of fluid per day) that more than one central battery is used.

Inlet Headers

To isolate individual parts of a gathering system, the system is often terminated in an inlet header. This header can be constructed of large pipe as

INCOMING LINES

LINES TO
BATTERY

Figure 1-18. Several satellite transfer lines or gathering system lines usually enter a central battery at an inlet header.

shown in Figure 1-18. Some gas gathering systems use the inlet header as a scrubber to remove liquid or stabilize slugging flow. The inlet header should be designed to be as large as possible to minimize friction losses.

Flow Splitters

Parallel process vessels are sometimes used with centralized processing facilities. One reason is that there may be too much fluid to process with a single, reasonable-sized vessel. Because all fluid is processed through a single facility, parallel vessels may be used so that if one requires service, the other vessels can process the fluid from the disabled vessel without requiring lease production to be curtailed. Although a header can gather fluid from several pipelines, it cannot easily distribute fluid to several vessels—even with extensive use of valves.

A flow splitter is a vessel that performs several functions. First, it functions as a two-phase separator. Second, internal baffles and long retention time allow the vessel to function as a three-phase separator in which free water is allowed to settle out of, and be removed from, the vessel

Figure 1-19. A flow splitter can distribute liquid among several treaters.

separately from oil and emulsion. Finally, the flow splitter has several compartments into which oil and emulsion flow. Individual floats allow the contents of these chambers to be released separately. Each float is independently adjusted, so the outlet flow rates from the compartments are not necessarily equal.

Figure 1-19 shows an application of a flow splitter using several treaters. The treaters are not all the same size and do not necessarily use the same inlet flow rates. Thus, the floats in the flow splitter are adjusted to give different rates to the treaters.

Auxiliary Heat Sources

In process facilities for some large leases some equipment may generate a great deal of heat which would normally be released to the atmosphere. Exhaust stacks from large engines are good examples. Heat used for treating is valuable because it normally requires the consumption of natural gas, so it makes sense to make some effort to tap this heat being vented to the air.

Heat exchangers (Figure 1-20) are pipe coils used to transfer heat from the surrounding exhaust gases to the fluid contained in the pipe. Heat

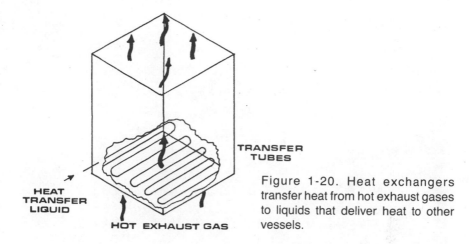

TRANSFER
TUBES

HEAT
TRANSFER
LIQUID

HOT EXHAUST GAS

Figure 1-20. Heat exchangers transfer heat from hot exhaust gases to liquids that deliver heat to other vessels.

exchangers capture and use waste heat. Figure 1-21 shows a facility using heat transfer from the exhaust of gas turbine engines to free-water knockouts. This particular facility uses no treaters because enough heat is recovered from the engines that additional heat is not needed.

In large facilities small heaters, such as those shown in Figure 1-22, are used to heat fluid without intending to actually treat the fluid at that time. These preheaters are similar to wellhead heaters discussed in Volume 2 (see Chapter 2, Surface Equipment for Flowing Wells).

Gun-Barrel Treating

Gun-barrel treating has long been an accepted method, and it may be used as the principal way to treat loose emulsions. More often, however, gun-barrel treating is used in conjunction with treaters in a loop of equipment. That is, emulsion may be heated, sent to a gun-barrel tank to stand for a time, then pumped to a treater where final treating is done.

Multiple-Pass Treating

To attain the highest level of treating efficiency in large batteries, several vessels and processes are often used to treat emulsion. Figure 1-23 is the flow diagram of a large central treating facility. Incoming fluid from the lease enters a gun-barrel tank that is heated by hot water from the treaters. Emulsion begins settling into oil and water layers in this tank. Emulsion is pumped from near the top of the emulsion layer to a treater. Taking emulsion at this level sends as little water to the treaters as possible. The treater then separates water and oil, sending the water back to the

HEAT TRANSFER
OIL LINE

Figure 1-21. Heat exchangers can trap waste heat for use in treating emulsions.

gun-barrel tank and oil to another tank. Water is removed from the gun-barrel tank by a pump operated with an interface float.

Optimization of Treating Techniques

Large processing facilities must be able to treat thousands of barrels of crude oil and sometimes tens of thousands of barrels of water in a day's time. When such volumes are involved, the treating efficiency becomes vitally important because massive quantities of fuel gas, demulsifiers, and electric power are required. Fortunately, large facilities process so much fluid that large initial investments in equipment are justified to attain maximum efficiency.

When treating efficiency is considered, not only is the cost of fuel, chemicals, and electricity important; the composition of treated oil must also be taken into account. When the treating temperature is greater than about 120° F, light hydrocarbons boil out of the oil and are collected with gas. The loss of these light components reduces the API gravity of the oil, resulting in reduced market value.

Figure 1-22. Small heaters are sometimes used to preheat fluid before allowing it to enter treaters.

The concept of multiple-pass treating is most effective in optimizing treating efficiency. Heat added in a treater is often reused in tanks or heat exchangers to preheat liquids. Preheating allows some liquid separation before the liquid reaches the treaters which, in turn, reduces the heating requirements of the treaters.

Using large tanks also assists in liquid separation. Incoming liquid is allowed to stand for quite some time and settle before treating ever begins. Even if some heat is added by recycling liquid, a settling tank is a very cost-efficient separation method. .

Electrostatic treaters are also cost efficient in treating emulsions. The use of grids allows the treaters to operate at low temperatures. Such treaters sometimes require no fuel gas at all. The cost of electricity to power the grids is small in comparison to the cost of gas.

Optimization of treating facilities is not always a simple task. Treaters which can effectively separate tight emulsions must sometimes be selected. These treaters must be incorporated into an overall facility that takes

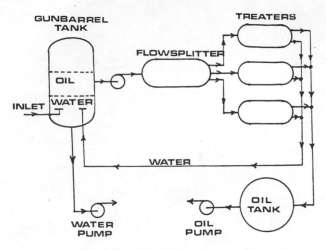

Figure 1-23. A gun-barrel tank may be used to exchange heat between liquids in a central battery.

advantage of the treaters, tanks, and other equipment. Chemical treatments may be required, but it is often necessary to wait until a facility is operational before determining the type and volume to be used. Finally, operators must be trained to make the equipment function as it was intended.

Storage

In principle, liquid storage in large central processing facilities is no different than in small facilities. In practice, however, storage in large facilities is complicated by the volumes involved, which may be several orders of magnitude larger than for small batteries. Instead of dealing with several hundred barrels of liquid which would require one or two 500-barrel tanks, a large battery can require storage volume of several thousand barrels, and many such small tanks are required.

Large tanks of 5000-20,000 barrels were at one time limited to large refining plants, but in the last few years many petroleum producers have incorporated these tanks into their systems. Figure 1-24 shows a 10,000-barrel tank used as the principal storage tank in a large process facility. This tank is obviously larger than the smaller tanks surrounding it, but the tank is also constructed differently than the smaller tanks. Large tanks such as this require heavy internal reinforcement of the walls and top deck to be able to handle the heavy forces exerted. The bases and the pads must also be of heavy construction to support weights of several million pounds.

Figure 1-24. Large tanks are commonly used as principal storage vessels in large central batteries.

Some batteries require storage of many hundreds of barrels but do not require such large tanks as shown previously. The facilities are designed to use several 500-barrel tanks or a similar arrangement. Such a system is pictured in Figure 1-25. This particular process had several groups of tanks connected together—equalized—so that each group serves a common purpose.

Lease Automatic Custody Transfer (LACT) Units

As the size of treating and storage facilities increases, so does the complexity of metering and transferring crude oil to the pipelines which carry it to refineries. Even when the transfer pipeline is owned by the same companies who produce the crude oil (this is not always the case) most states have strict regulations about metering, volume, and composition of oil as it enters the pipeline. With small production facilities of only a few oil storage tanks and less than a few hundred barrels daily production, metering and sampling of oil is done with tank gauge lines and tank thiefs discussed in Volume 2 (see Chapter 7, Level and Sample Measurement). However, when a lease uses many small tanks or very large tanks to handle several hundred barrels of oil production daily, this manual method of metering and analyzing oil is not practical.

A Lease Automatic Custody Transfer Unit (LACT) is used not only to meter but also to analyze the oil transferred from a producing lease to a pipeline. A LACT unit (shown in Figure 1-26) is composed of a basic

Figure 1-25. Several small tanks may be used for storage in a large central battery.

sediment and water (BS & W) monitor (See p. 24), a diverting valve, a precise liquid meter (usually containing provisions for correcting the volume measured for the expansion of oil due to pressure and temperature variations), an automatic sampler, and a sample container. Oil, pumped by separate oil-transfer pump(s), passes first through the BS & W monitor, which monitors the water content of the oil. If the water content is much more than about 0.1%, the BS & W monitor signals the diverting valve to send the oil back to the production facility as unacceptable.

If the water content is low enough to meet pipeline standards, the valve allows oil to pass to the meter and sampler. Warm oil occupies slightly more volume than cool oil, and oil under high pressure occupies slightly less volume than oil under low pressure. To make the volume measurement fair to both the producer and the pipeline owner, the meter automatically corrects the volume to standard conditions (usually 60° F and sea level atmospheric pressure), and the metered volume is adjusted slightly to compensate for different conditions.

Figure 1-26. A Lease Automatic Custody Transfer (LACT) unit is usually the final transfer point from a central battery to a pipeline.

The sampler is operated by the meter and functions by allowing a few drops of oil to enter the sample chamber for every barrel or every 10 barrels passing through the meter. After some time, the sample is collected and analyzed for water content. This analysis gives the average water cut over a long period of time. The metered volume is then corrected by the few tenths of 1% of water that was metered as oil.

After oil has passed through the LACT unit, it enters and is transported in the pipeline. A LACT unit is a fully automatic device. Periodically, the sample container is analyzed. Also, the meter is checked for accuracy with a meter prover, which injects a precise volume of liquid through the meter. The meter's reading is compared to the known volume, and the meter is adjusted as needed or the readings from the meter are adjusted as required. Many producers prefer to read the meter and check the functions of the BS & W monitor periodically, but the entire unit is intended to operate unattended for weeks at a time. In fact, all valves are sealed with labeled lead slugs to assure that the valves are not switched, and the automatic operation of the unit is not compromised.

Figure 1-27. A capacitance probe is the sensing element that determines the amount of water in oil.

BS & W Monitors

A BS & W monitor or probe is an electronic device intended to measure the amount of water contained in an oil-water emulsion. When tank thiefs were the only method of sampling the contents of a tank, the resulting impurities were called basic sediment and water, BS & W, or just BS. This is the origin of the name of the device, although it was never intended to measure anything but water.

The most common BS & W device is a capacitance probe shown in Figure 1-27. An electronics circuit using this device as one component gives an electrical signal as the capacitance of the probe changes. This capacitance changes when even minute amounts of water are contained in an emulsion.

When they were first introduced, BS & W monitors were intended only for use with LACT units and thus measured emulsions containing at least 99% oil. Later, it was found that similar devices could be used for well testing where probes could be expected to measure emulsions containing 100% oil down to less than 1% oil.

Coordination of LACT Units with Other Equipment

LACT units are often used when central treating facilities are employed. Most operators consider them part of the treating facility and incorporate their functions into the operation of the battery. For example, the LACT units must be installed so that their operation complements that of the storage tanks. To avoid excessive temperature correction in the LACT meter, many operators prefer to leave liquid in storage as long as practical to allow it to cool before entering the meter. Storage tanks are often intentionally elevated above the level of the LACT unit so there is no possibility of gas entering the meter and interfering with its operation.

LACT units are integral parts of batteries using multiple-pass treating. When the LACT unit starts by-passing "bad oil" (oil containing too much water), the oil is sent to recycle tanks or treaters where it is joined by other liquid which is to pass back through the treaters. Figure 1-28 is a flow diagram of a battery using a LACT unit as part of the treating system.

Pipeline Transfer Pumps

The pumps used to move oil from tanks to pipelines are centrifugal or positive displacement—depending on the volume and pressure of the pipeline (See Volume 2, Chapter 7, Liquid Pumps). Figure 1-29 shows several centrifugal pumps used to pump large volumes of oil at low pressure. The reason for several pumps is that oil is not pumped at the same rate throughout a day, and at some times only one pump is needed, while at other times all must be running. Figure 1-30 shows a large positive displacement pump used to pump oil at high rates and high pressures. Because oil flows into and from this battery at a steady rate, one or two pumps are used and are expected to run continuously.

Operation and Maintenance

The equipment used for major production facilities is by far the most visible part of any production system. But this equipment is not necessarily the only feature of the system which affects the efficiency and cost of processing oil. Equipment operation must be made simple and straightforward—not because personnel might have problems with it—but because at times when something goes wrong, it is important that equipment be operated and adjusted quickly and easily. It is during the design and installation of equipment that later operating procedures can be made simple or difficult by the way in which equipment is established. Obviously, every effort should be made to simplify equipment operation when the equipment is installed.

Every piece of equipment, no matter how well constructed, will eventually fail in some way. When failures occur, the equipment must be repaired before it can be placed back in service. Operating procedures can sometimes prevent or delay failures. Preventive maintenance activities, such as frequent and careful inspection and lubrication, can extend the life of

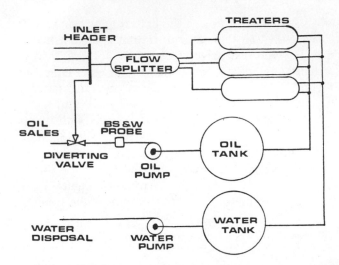

Figure 1-28. A LACT unit can recirculate oil in a central battery if the water content is too high.

Figure 1-29. Centrifugal pumps may be used to transfer liquid in a central battery.

Figure 1-30. Positive-displacement pumps are used for pipeline transfer when high pressures are involved.

most equipment. Even when inspection reveals that something is on the verge of breaking, the actual repair can sometimes be delayed until a convenient time when the equipment load or throughput can be reduced or by-passed.

Equilibrium in Treating Facilities

When a major treating facility is operating normally, it develops a kind of equilibrium in which the tank levels are fairly steady or at least stay within a well-defined set of levels, the treaters hold liquid at stable levels and produce oil with nearly constant water cut, pumps operate at regular intervals, and other equipment performs predictably. As long as the battery stays in this equilibrium condition, a major facility can often operate with minimum observation and intervention. However, a sudden change in operating conditions such as an equipment failure, a surge in inlet fluid flow rate, or a major adjustment in operating conditions can upset this equilibrium. For the next several hours, close attention and manual operation must be performed on the system to regain equilibrium (get the system "lined out"). Again, by carefully selecting and installing the pieces of equipment used in the facility, operators can minimize the difficulty of regaining equilibrium after an upset.

Commingling

Most state oil and gas regulations require that oil produced from different reservoirs (usually several formations at different depths) be kept separate until custody transfer, although the reservoirs may be contiguous and the individual wellbores are adjacent to each other. When major centralized treating facilities are involved, this separation requirement can mean that much of the expensive equipment must be duplicated. On such occasions, it is sometimes to the producer's advantage to obtain permission to commingle the oil streams together. This procedure may require additional equipment such as meters, mixing vessels, and piping that might not otherwise be used. However, the additional equipment may be justified if large items such as treaters and tanks can be deleted.

Water Production

This chapter has discussed the various methods of separating water and oil so the oil can be sent to refineries for further processing. However, no comments have been made about the disposition of water produced with oil. Chapter 2 discusses water handling and disposal.

Chapter 2

Water Handling
and Treating

Virtually every reservoir contains some water with hydrocarbons, and this water usually comes to the surface with oil and gas. Produced water has dissolved salt in it because the water passed through salt formations before entering the petroleum reservoir. Gas and oil are refined to separate their constituent parts into usable products in processes which cannot function with more than traces of water in the streams. Thus, water must be removed from oil and gas before they can be transported to refining facilities. Because the water contains at least salts and probably chemical compounds that would be dangerous to life, water must be carefully disposed of after separation from hydrocarbons.

Final Water Treatment

Water is usually separated from oil in treaters or tanks, but a little oil may be left with the water. In major treating facilities large volumes of water are often handled. When as little as 1% of this volume is oil, simple disposal of water can result in the loss of sizable volumes of oil. Not only can the oil be lost, but it can have an adverse effect on disposal wells, pumps, and other equipment. Thus, it is important that all oil be removed from water before it is pumped to disposal facilities.

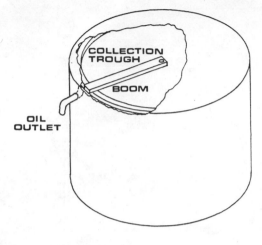

Figure 2-1. A rotating boom pushes oil into a trough from which it drains back to oil-treating equipment.

Settling Pits

Water is sometimes pumped into shallow pits where it is allowed to stand quietly. Entrained oil droplets usually settle out of the water and coalesce into a layer of oil floating on the water. This method is not altogether satisfactory and is not used as often as other methods in new facilities. Although the pits must be lined with something such as plastic, asphalt, or concrete, water usually seeps into the ground and can contaminate shallow fresh-water aquifers or damage cropland. When exposed to the atmosphere, oil can evaporate and be lost. Extensive fencing systems are required to keep livestock and youngsters away from the pits. Although the water may not necessarily be dirty, it is almost invariably poisonous. With these problems in mind, most producers have opted for different methods of final water treatment.

Skimmers

Water can be pumped into tanks equipped with rotating booms called skimmers (Figure 2-1). As the water stands quietly, oil separates and rises to the surface of the water. The rotating booms push the oil into collection troughs on the sides of the tank from which the oil can be pumped back to processing facilities and recovered. The tanks used for this method may be closed or open-top. Although open-top tanks are easier to service, the same evaporation problems can occur as in open pits. Closed tanks are usually preferred to avoid oil evaporation and air pollution.

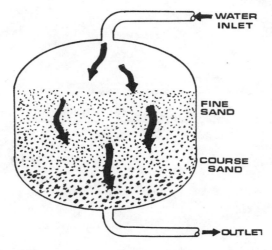

Figure 2-2. Sand filters trap solids before they enter an injection system.

Filters

Water to be disposed of often contains all the solids that were originally produced and passed through various production equipment. To keep these solids from plugging disposal equipment, filters are often installed. Sand filters (Figure 2-2) are steel vessels filled with sand. The sand is graded from course grains (where the water enters) to fine grains (where it leaves) to assure that all particles are removed. These sand filters are somewhat effective in removing oil as well as solids from water. The multiple surfaces of the sand grains offer places for the oil droplets to coalesce and stop in the filter. Oil and solids eventually plug the sand and must be removed. This is accomplished in an operation called *backwashing* in which the flow of water through the filter is reversed and forced to a tank where the solids can be collected and the oil recovered.

Coalescers

Coalescers are closed vessels (Figure 2-3) filled with hay or excelsior through which water is passed. Entrained oil is attracted to and collects—coalesces—into large drops on this material and stays within the vessel, while clean water passes through the vessel. The fiberous material also traps and collects some solid particles that may be carried by the water. A coalescer does not remove solids as efficiently as a sand filter. Periodically, coalescers must be backwashed to remove trapped oil and solids.

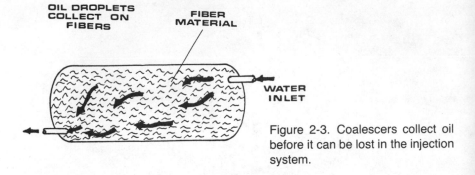

Figure 2-3. Coalescers collect oil before it can be lost in the injection system.

Aerated Skimmers

Another water treatment system, an aerated skimmer, is shown in Figure 2-4. Incoming water is agitated and aerated with a gas such as carbon dioxide, nitrogen, or sometimes air that does not interfere with production operations. This agitation with gas causes tiny droplets of entrained oil to coalesce into large droplets that are literally pushed to the top of the skimmer by the agitation of the liquid. Paddles rotate across the surface of the water and push the oil into collection troughs for removal to treating facilities. Since this is a closed vessel, there are no pollution problems with the device. Also, these vessels are much more efficient in removing oil than tank skimmers.

Storage Facilities

When water is produced in a large treating facility, some care must be given to selecting the method of storing the water before it is pumped away. Small tanks may be used for small volumes, but when the volume dictates, much larger tanks may be required. However, because of the special corrosion problems with water storage, very large tanks are not often used for storage of large volumes of water. Water is a natural oxidizing agent for most metals, particularly alloys of iron. This problem is aggravated when water contains dissolved oxygen. Corrosion of steel tanks can occur very rapidly and cause leaks that must be stopped before pollution problems can develop. Many operators opt for several small water tanks rather than a single, large tank so that if a leak develops, that tank can be taken out of service for repair while water is sent through the other tanks without interrupting production operations.

Water tanks require more corrosion protection than tanks used for most other types of service. A thick coat of plastic or epoxy is usually applied to all

Figure 2-4. An aerated skimmer uses bubbling gas to coalesce oil droplets so skimmer booms can collect the oil into troughs inside the skimmer.

inside surfaces of water tanks. This coating is carefully inspected for holidays to be sure the coating is adequate. Even when a reliable coating is used, water tanks are periodically taken out of service, drained, cleaned, and carefully inspected for signs of corrosion.

Pumping Facilities

Water is usually pumped to whatever disposal facilities are used for a lease. If water is merely pumped from a tank to another tank or to open pits, low-pressure, high-rate centrifugal pumps may be used. When water is pumped to disposal or water injection wells, high-pressure pumps are usually used. These pumps may be the positive displacement type (Figure 2-5), much like those used for pumping oil into pipelines. High-pressure centrifugal pumps (Figure 2-6) are also used for moving produced water, although these pumps are usually reserved for the highest-volume applications. Centrifugal pumps are particularly effective when large volumes of solid particles are pumped because most centrifugal pumps utilize parts with a great deal of clearance and do not wear as much as positive displacement pumps.

Figure 2-5. High-pressure positive displacement pumps move liquid in injection systems.

Water pumps must be equipped for water service because of the inherent corrosion problems. Stainless steels or other alloys which are corrosion-resistant are used for plungers, valves, impellers, and sometimes pump cases. The pipe connecting these pumps to other vessels is usually internally coated with plastic to minimize corrosion problems.

Disposal pumps are usually equipped with both suction and discharge pressure desurgers. The suction desurgers are required because pumps are often connected to storage tanks with rigid pipe, and this high-volume application can cause fatigue failures in rigid pipe. Discharge pressure desurgers (Figure 2-7) are used because the outlet of a disposal pump is usually high pressure and high rate. Such operation usually causes fatigue failures in discharge piping without desurgers.

Saltwater Disposal

Water can be disposed of in one of several ways. For many years, the common method of saltwater disposal was to place the water in open pits. Evaporation and percolation into the earth's surface removed the water fairly quickly. Open-pit disposal has been largely abandoned because of the

Figure 2-6. High-pressure centrifugal pumps are also used as water injection pumps.

possibility of danger to human, animal, and plant life. Also, when salt water leaks into the earth's surface, it almost invariably travels to shallow fresh-water aquifers where it contaminates water used for domestic and industrial consumption and irrigation.

Open-pit disposal has been largely replaced by saltwater disposal wells. Some operators contract with services which haul water away from the lease for a fee. These services usually own one or more saltwater disposal wells and the associated facilities—not associated with the production lease—into which the water is pumped.

Saltwater Disposal (SWD) Wells

Saltwater disposal wells are usually located outside (or near the outside edges of) large leases because it is best to dispose of water as far from the heart of the reservoir as possible to avoid excessive water injection if a casing leak ever develops. Thus, SWD facilities are usually remote from other lease operations. SWD facilities are usually equipped with tanks and pumps

Figure 2-7. Desurgers are usually required to remove pressure fluctuations from positive displacement pump outlets.

located close to the wellheads. Figure 2-8 is a photograph of a SWD facility located some five miles from the lease it serves. Water is pumped to this facility from central treating facilities and temporarily stored in the tanks, while the pumps inject the water into an aquifer under the productive reservoir.

The subsurface equipment of an SWD well is usually arranged as shown in Figure 2-9. A packer isolates the casing from the high pressure used to inject water at high rates. Tubing (coated internally and externally with plastic) must be strong enough to withstand the high pressure. The casing is usually perforated extensively to allow free passage of water.

Metal corrosion is one of the major problems with most SWD wells and their associated storage and pumping equipment. Unlike other petroleum production equipment, SWD equipment seldom contacts any liquid but very corrosive water. Water to be disposed usually contains large amounts of dissolved air. It also contains spent acid water and other chemicals used in production and stimulation work throughout a lease. Thus, the liquid contacting the equipment in an SWD facility is most likely to cause severe corrosion unless care is exercised to prevent it.

Subsurface equipment in an SWD well is usually minimized so that little protection is required for the equipment. However, an SWD well requires at least a packer, tubing, and casing. The packer is usually constructed of special steel alloys which resist the attack of corrosive liquids.

Figure 2-8. A saltwater disposal station usually consists of storage tanks and high-pressure pumps.

The tubing used in an SWD well is the same used in other wells and must be protected from the corrosive environment. Internal and external plastic coating is of some benefit to the tubing, but eventually it will fail to protect the metal. Tubing must be removed and replaced when the plastic coating breaks down.

Casing cannot be coated in the well and is left bare in SWD wells. To avoid leaks, casing must be protected indefinitely and thoroughly.

The inside of casing and the outside of tubing is protected simultaneously by liquid in the tubing-casing annulus above the packer. Water, mixed with a large amount of corrosion inhibitor, is dumped into the annulus and left there undisturbed. This inhibitor mixture is usually effective in protecting the tubing and casing from corrosion as long as the packer effectively seals out pressure below it. Most operators regularly monitor the casing pressure of SWD wells because this pressure begins to increase when a packer leak develops. It is important that such a leak be corrected upon discovery and the corrosion inhibitor mixture replaced immediately.

Casing below the packer is seldom protected. First, there is no realistic way to do it, and second, there would be few problems even if this section of casing corroded completely. SWD casing is usually "blanket perforated" (perforations concentrated as heavily as possible by making multiple runs with a perforating gun). If more holes develop as a result of corrosion, no

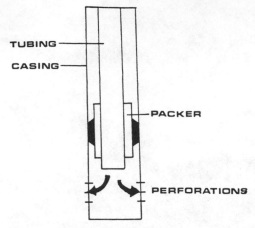

Figure 2-9. The subsurface equipment in a SWD well usually consists of tubing and a packer to isolate the high bottom-hole pressure from the casing.

harm is done. Thus, few producers make any provisions for protecting the lower casing.

Scale Accumulation Caused by Saltwater Disposal

Although the water being injected into an aquifer may have been produced originally from the reservoir immediately above it, water injected into an aquifer can cause scales—salts that result from reactions of various chemicals in dissimilar waters—to accumulate in the wellbore and in the formation. Scales form when water being injected is incompatible with the water already in the aquifer. That is, the two waters contain chemicals that can react and leave salts as residues. Because water to be disposed of is a mixture of water, treating chemicals, and the residues of processing, it is not likely to be compatible with even its own neighboring water from the reservoir. Little short of extensive chemical processing can be done to make disposal water compatible with other waters. Even if there were a feasible method, the cost of such treatment could not be justified. Most operators expect SWD wells to plug eventually as a result of scale accumulation. When they do, the operator may acidize the well to remove the scale, by-pass the blocked zone, or opt to abandon the well altogether and convert another well to SWD service.

Use of Produced Water for Waterflood Applications

Secondary and improved recovery systems usually require water injection for at least part of the recovery mechanism. Many operators have found that produced water can be used quite well for injection. This provides not only a partial source of injection water but also an alternate means of disposing of produced water. Some water must be disposed of in SWD wells even when the majority of it is being used for injection; however, the volume of water to be disposed of is reduced drastically. The produced water must be treated for injection or at least sampled periodically to be sure it does not cause scaling problems in injection facilities. Mixing of waters and the attendant problems, as well as water injection for recovery purposes, are the subjects of Chapter 3.

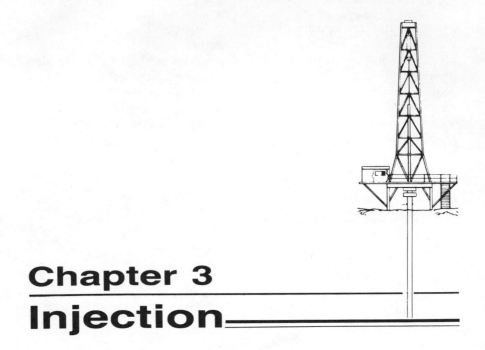

Chapter 3
Injection

Water Injection

Since the mid-1950s, water injection for secondary recovery has been the dominant method of supplying energy to a reservoir to assist reservoir pressure in pushing fluids to production wells. Since the early 1970s, water injection has also played an important part in virtually all enhanced recovery techniques.

Unlike water being pumped into an aquifer simply to get rid of it, water pumped into a reservoir for recovery purposes is injected under very carefully controlled conditions. It is pumped into a reservoir slowly—several hundred barrels daily—at carefully monitored and controlled surface pressure to avoid fracturing the formation or distorting the pattern of flood banks.

Water injection is occasionally used on small leases with fewer than 50 wells, but usually this recovery technique is reserved for larger projects. Injection projects usually consist of a number of injection wells arranged in patterns (see Volume 1, pp. 34-36). Figure 3-1 shows a "five-spot pattern" of injection and production wells (so named for its correlation to the pattern used on dominoes and dice), while Figure 3-2 shows a "line-drive pattern" of wells. For these orderly patterns to build effective flood banks, a great number of injection and production wells are required.

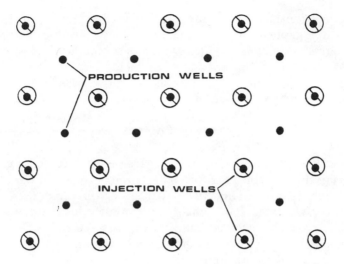

Figure 3-1. The five-spot injection pattern is a common pattern.

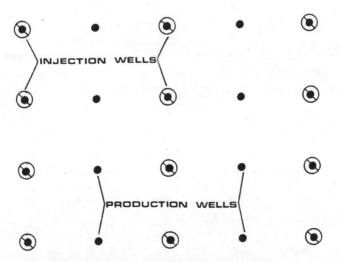

Figure 3-2. The line-drive pattern, where injection wells are arranged in rows, is another injection pattern.

Water injection facilities require a great deal of equipment. Wells must be chosen and equipped for injection. Then pumps must be placed to give enough pressure, and distribution systems must be installed to provide a means of transporting water to the wells. However, the most important initial consideration in establishing a waterflood system is securing an abundant and reliable source of clean water.

Water Sources

Produced Water. Water produced from a lease with hydrocarbons may be used for injection into the reservoir. In fact, water produced from a reservoir is the preferred source for injection into the reservoir. If produced water has not been contaminated, it is probably compatible with the water still in the reservoir. Produced water is certainly the cheapest source of water, since not only is it free, but it is an expense item if not used for reinjection. Of course, to be used for injection, produced water must be clean and completely oil-free.

Fresh Water. In some areas fresh water is available in abundant quantities from shallow fresh-water aquifers or from rivers and lakes. When it is available, fresh water is usually an excellent source for injection water. Because this water is near to or at the earth's surface, it contains significant amounts of dissolved oxygen. This oxygen encourages corrosion in steel equipment and requires some special treatment. In addition, fresh water usually contains bacteria which, in sufficient numbers, can plug equipment used in injection systems. Another consideration when using fresh water for injection is that this is the same water that is usually used for domestic and industrial consumption and for irrigation. Waterflood projects do not use water as rapidly as most irrigation operations. For example, a typical small farm of 640 acres might use as much as 6400 acre-feet (a volume with a surface area of 640 acres and a depth of 10 feet—equivalent to more than 270,000,000 cubic feet or more than 49,000,000 barrels) of water for irrigation in a growing season, while a rather large water injection project might require 100,000 barrels of fresh water per day (equivalent to slightly more than 200,000,000 cubic feet or about 4700 acre-feet per year). The 5-10 farms that could occupy the surface under which a large waterflood project was operated could easily use 5-10 times the water used by the production operation. Of course, petroleum leases could not use water from an aquifer if fresh water for domestic consumption and irrigation were in short supply as is the case in many semiarid areas.

Outside Saltwater Sources. Salt water can sometimes be secured for use in injection systems from outside sources. Leases in coastal areas, of course,

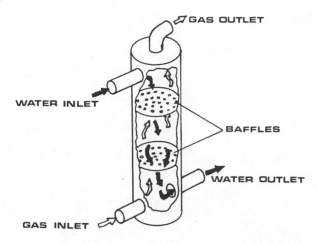

Figure 3-3. Natural gas bubbling up through water removes oxygen from the water in a water deoxygenation tower.

have the largest water body—the ocean—as a source. Some large leases produce so much more salt water than needed for injection that excess water is sold to outside leases. Outside saltwater sources can pose some of the same problems as fresh-water sources. First, the water can contain dissolved oxygen, a source of corrosion problems. Second, it can contain bacteria which can plug equipment and formations and chemicals that make it incompatible with formation water. Of course, the problem of competition with domestic and irrigation water sources disappears.

Oxygen Removal

Oxygen reacts with iron in steel to produce oxides—commonly called rust—in an oxidation reaction similar to combustion which releases so little heat that the temperature does not rise. When water contains dissolved oxygen, the problem is compounded as witnessed by the corrosion of steel left outside for any length of time.

Oxygen can be removed from water in a process called *deoxygenation*. In petroleum production technology water deoxygenation can be done mechanically or chemically. Figure 3-3 is a cutaway view of a water deoxygenation tower commonly used in water processing facilities. Natural gas is injected into the bottom of the tower, while water enters from the top. As the water is forced to contact the gas by flowing through numerous trays and baffles, the gas absorbs oxygen from the water. Water containing little oxygen (five parts per million—ppm) leaves the vessel through the bottom, while oxygen-laden gas leaves through the top. Although the gas contains some oxygen, it is not particularly flammable, but it contains so much water

Figure 3-4. Water deoxygenation towers are commonly used on large water injection stations.

vapor that a dehydrator is required to clean the gas for processing. Many operators avoid this problem in large central processing facilities by using the gas from deoxygenation towers as treater and heater fuel—an excellent example of optimization of facilities in a system. Figure 3-4 is a photograph of a water deoxygenation tower used in a large central facility for processing both oil and water.

Oxygen is also removed from water using chemicals called *scavangers*. These chemicals are injected into a stream of water, allowed to mix with the water, and react with the oxygen. The reaction converts oxygen from its gaseous form to a liquid compound that is inert to reactions with steel or other materials involved in injection. Chemical deoxygenation is usually expensive because of the cost of the scavangers, but no special equipment is required besides chemical pumps. Also, chemical deoxygenation does a better job of removing oxygen, usually leaving one ppm or less. Since some bacteria can survive and oxidation can occur with oxygen concentrations in the range of 5-10 ppm, it is necessary to remove oxygen effectively.

Water-Mixing Vessels

Sometimes waters from several sources are used for injection. For example, produced water and fresh water are used to compose the supply of

Figure 3-5. Water-mixing vessels assure even blending of several water streams.

injection water. To assure that deoxygenation or corrosion treatments act on all water involved, a special vessel may be required to assure that all waters are thoroughly mixed. Figure 3-5 is a photograph of a water-mixing vessel. This vessel contains a series of baffles that cause several water streams to mingle thoroughly by flowing through serpentine paths. Mixing vessels are not always necessary, but they are sometimes useful when chemical treating is to be done.

Bacteria Treatment

Bacteria is a prevalent problem in many water injection systems. This life does not damage equipment, but when bacteria concentration becomes high enough, it can plug small passages in equipment and in the formation.

There are several kinds of bacteria that affect injection systems. Aerobic bacteria requires five ppm of oxygen or more for survival and can be eliminated effectively by reducing the oxygen content to less than five ppm. Anaerobic bacteria derive their oxygen from water molecules and require no dissolved oxygen.

Bacteria are removed from water streams by chemical treatment with bacteriacides. One common solid-bacteria treating chemical is sodium hypochlorite, the chemical often used to chlorinate swimming pools. Other solid and liquid chemicals are also available to treat water for bacteria. Liquids are simply pumped into a water stream, while solids are first mixed with water and then pumped into water streams. Bacteria control is maintained with a sufficient concentration of chemicals in the water. When a great concentration of bacteria has accumulated, shock treatments are required to reduce the bacteria concentration so that low-concentration treatments can maintain control. Essentially, shock treatments consist of applying large volumes of highly concentrated bacteriacide for short periods of time. Once the bacteria population is reduced, regular treatments can maintain a low population of bacteria.

Scale Treatments

When water is injected into a formation, there is the possibility that compounds in the injected water will react with compounds in the formation water. This reaction creates salts which precipitate as scale. Before an injection system is started, samples of the formation and injection waters are mixed, and their scaling tendencies (the likelihood that scales will form and precipitate) are measured. When there is little possibility that scales will form, the waters are said to be compatible. If it is likely that scale deposits will result from mixing them, waters are said to be incompatible.

Even when waters are compatible, there is the possibility that some scale deposits will develop in subsurface equipment and in the formation. Scale inhibitors are usually mixed with injection water to reduce scale accumulations; however, even then some deposits can develop. Some injection wells must be stimulated occasionally to remove scale deposits. Scale is usually removed with acid, but an injection well must be acidized very carefully to keep from fracturing the formation and distorting the pattern of injection in the future. Scale inhibitor squeeze treatments are sometimes effective in providing long-term protection against scale accumulations, and continuous treatments may also help in alleviating scale problems.

Water Injection Pumps

Pressured water for injection is supplied by pumps similar to those used for simple transportation. Most water injection systems require high pressure (in the range of 1000 psi) at the wellheads so that the combined hydrostatic head of the water and the surface injection pressure is high enough to push water into the formation. Because there are many injection wells in most injection systems, large volumes of water are required. Therefore, injection pumps must be large-volume, high-pressure pumps able to serve the injection systems. Because so much water is moved and the horsepower requirements of the pumps are large, it is important that injection pumps be very efficient. Even small losses in efficiency (a few percent) can be sizable in terms of the energy requirements of the prime movers.

Positive Displacement Pumps. The positive displacement-type pumps discussed in Volume 2, Chapter 7 are often used for water injection. In

Figure 3-6. Several positive displacement pumps are usually used in water injection stations.

general, positive displacement pumps are less expensive and more reliable than other pump types available. On the other hand, they are usually available only in a limited range of pumping rates. Figure 3-6 shows several quintuplex (five-plunger) positive displacement pumps operating in parallel. Pumps operating in parallel all take suction from a common pipe and discharge into a common pipe or header. (The pumps all have the same discharge pressure, but the rate from the header is the sum of the rates of all the pumps.)

Split-Case Centrifugal Pumps. These are a specialized type of centrifugal pump used for large-volume, high-pressure applications. They operate as a staged pump, much like the centrifugal pump in an electric submersible pumping system—each stage raises the pressure incrementally and sends the water on to the next stage. Split-case centrifugal pumps are used with rates of 10,000-50,000 barrels per day (300-1500 gpm) at pressures of 1000-2000 psi. They are expensive and somewhat delicate but are ideally suited for pumping water into an injection system. If the total injection rate is slowed for some reason, the water simply slips past the pump impellers instead of being forced through regardless of pressure (as is the case with positive displacement pumps). Figure 3-7 is a water injection station which pumps approximately 60,000 barrels per day at about 1000 psi by operating several pumps in parallel. Even with large positive displacement pumps, a number of pumps are required to do the same job these pumps do.

Centrifugal Pumps. Centrifugal pumps of the same type discussed in Volume 2, Chapter 7 are sometimes used as injection pumps in waterflood projects where low surface pressures are required. These pumps are usually designed for low discharge pressure (100-300 psi) and are not usually well suited to high-pressure water injection systems. However, centrifugal pumps are very well suited to large-volume, low-pressure applications, and they are ideal as transfer pumps within an injection facility or as charge pumps for other high-pressure pumps. In other words, low-pressure centrifugal pumps are sometimes used to pump water at high rates and fairly low pressure (50-100 psi) directly into the suction line of a high-pressure pump, which raises the pressure to the required injection pressure.

Desurgers. As is the case with other high-volume or high-pressure pumps, injection pumps generally require suction and discharge desurgers. Desurgers become more important as the discharge pressure and rate increase. Of course, the size and pressure limitations of desurgers also vary with rate and pressure. Suction desurgers become critically important in injection stations because the high rates involved usually require large pipe which is inherently weaker than comparably sized small pipe. The large-diameter, low-pressure pipe used for suction lines and headers is usually fairly weak; desurgers are required to avoid fatigue damage to the pipe. Some operators connect the pump suction lines to headers or tanks with large, strong hoses so the hoses can move slightly without damaging the pipe.

Prime Movers for Injection Pumps and Compressors

For a pump to move liquid or a compressor to move gas (discussed later in this chapter), the drive shaft must be turned by a prime mover of some sort. Pumping fluids at high rates and high discharge pressures requires a great deal of power from the prime mover. When liquid is pumped, the power in horsepower is 0.00011 times the product of the rate (in gpm) and the difference in pressure (in psi) between the suction and discharge. For example, to pump water at 1000 gpm (about 34,200 barrels per day) and raise the pressure from 100 to 1000 psi, 525 horsepower is required, assuming 100% pump efficiency.

Figure 3-7. Several centrifugal pumps can serve as primary pumps in an injection station.

Several means are available to drive injection pumps and compressors. These are usually very large machines and may range from several hundred to more than 1000 horsepower. Such large prime movers are sometimes delicate simply because of their size. Regular preventive maintenance is required to assure that a problem does not develop. Electric motors, turbine engines, and gas engines are available for use as large prime movers.

Electric Motors. Although some injection pumps are small enough to require conventionally sized motors of less than 150 horsepower, most injection applications of electric motors require large motors of 200 horsepower or more. Electric motors of less than 200 horsepower are driven by electric current at 480 volts, but larger motors are usually driven with higher voltage (2400 or 4160 volts). The starters and electric equipment external to the motor are much larger and more heavily protected against accidental entry than the 480-volt equipment used for smaller motors. The motors are not necessarily more delicate than smaller motors, but they are more carefully protected against overloading and improper operation than small motors. Large motors are usually expensive and repair of a damaged motor can be very time-consuming and costly, so more precautions are

Figure 3-8. Large centrifugal pumps may be driven by large, high-voltage electric motors.

necessary to avoid damaging them. Figure 3-8 is a photograph of a 1000-horsepower, 2400-volt electric motor used to drive a large, split-case centrifugal pump in a water injection station. Figure 3-9 is the electric switching equipment necessary for the motor to operate. Figure 3-10 is a 200-horsepower, 480-volt electric motor used to drive a positive displacement pump. One advantage of electric motors in injection service is that they can be started and stopped automatically without an operator standing by, although some companies make it a practice to have someone available when a large pump is placed in or taken out of active service.

Gas Engines. Gas engines are also used for injection operations. Such engines are available in sizes up to several thousand horsepower, and they are usually durable and dependable. This dependability should never overshadow the need for regular, periodic maintenance. Figure 3-11 is a large gas engine driving a split-case centrifugal pump. Figure 3-12 shows the operating panel for a large gas engine. The gauges show such vital parameters as engine temperature, oil pressure and temperature, and suction and discharge pressure. When emergency conditions such as engine malfunctions or high discharge or low suction pressure occur, the small switches (called *tattletales*) pop out, short out the ignition system, stop the

Figure 3-9. High-voltage electric motors require specialized switching equipment called starters.

engine, and close fuel valves and suction and discharge valves. A control
panel can be built to annunciate the problem and control the engine for
virtually any condition imaginable.

Gas Turbine Engines. Gas turbine engines are rapidly becoming a common
way of driving large horsepower loads such as injection pumps and
compressors. There can be no doubt that gas turbine engines are much more
complex than either gas engines or electric motors, but turbine engines are
efficient in driving heavy loads, and this efficiency justifies the complexity.
Figure 3-13 is a cutaway diagram of one such turbine. The mixture of fuel
and air burns in a jetting manner, and the rapidly expanding exhaust gas
turns two sets of shafts: one rotates the compressor that drives combustion

Figure 3-10. Small electric motors are commonly used in water injection stations.

air into the combustion chamber; the other turns the output shaft to which the load is coupled. Actually, the operation of a turbine engine is far more complex than can be shown in a simplified diagram because a number of systems (such as starting, lubrication, and control systems) are involved. Figure 3-14 shows a gas turbine engine. The complex electrical, electronic, pneumatic, and hydraulic systems used to start the engine, lubricate the rapidly moving parts, and control the output speed and torque can be seen in this figure. Figure 3-15 shows the interior of a control station for a turbine engine. Figure 3-16 shows an injection station using several gas turbine engine/pump combinations to pump more than 100,000 barrels per day at a better than 1000-psi discharge pressure. Because these units are outdoors, the engines are protected in weatherproof enclosures.

Heat Exchange in Large Gas Engines

Although burning natural gas as fuel for internal combustion and turbine engines is a common method of driving large loads, it is not a particularly efficient means of energy utilization. After natural gas has been burned and the expanding exhaust gases are used to drive the output shaft through turbine impellers or pistons, the exhaust gases are still hot at more than 500° F. Although these hot gases are channeled upward so they cause no environmental problems, the loss of heat represents energy inefficiency. In

Figure 3-11. Large pumps may be driven by gas engines.

an effort to use the energy liberated from burning gases effectively, many producers have installed equipment to trap the waste heat from exhaust gases with heat exchangers. Heat exchangers simply pass hot gases past tube bundles containing liquid which is heated by the gases. This hot liquid is then pumped to treating equipment where it heats emulsions, drives steam turbines, or is used in any other application needing heat.

Figure 3-17 shows a heat exchanger that traps the waste heat from a gas turbine engine. Heat-transfer oil (much like a light lubricating oil) is heated in the exchanger and pumped through a closed circulation system (Figure 3-18) to free-water knockouts where it heats emulsion for crude oil processing. In this system it is important that the heat-transfer oil remain at a fairly constant temperature; therefore, the heat exchangers are equipped with pneumatically operated baffles that allow atmospheric air to enter and cool the exhaust gas if the oil temperature rises too much. Also, this system uses waste heat to preheat combustion air and fuel gas to improve energy conversion efficiency during combustion. All in all, this particular system is quite efficient in the use of the energy released to operate the turbines; it is

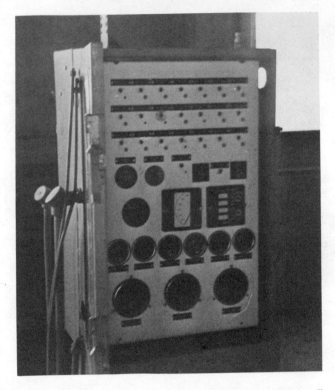

Figure 3-12. The operating panel for a large gas engine shows all necessary operating information about the engine and provides protection against engine malfunction.

the only natural gas usage in a facility that can process more than 30,000 barrels of crude oil daily and inject more than 100,000 barrels of water daily at more than 1000 psi.

Gas Injection Facilities

Gas injection for pressure maintenance (discussed in Volume 2) had largely been replaced by water injection and, for a time, it appeared that gas injection had become a thing of the past. However, with the advent of some enhanced recovery methods, gas injection has become a vitally important area of production technology. Gone are the days, however, when gas was simply compressed and injected into wells at whatever pressure was required to maintain a certain pressure in the reservoir. Rather, gas mixtures must be very carefully processed, blended, and compressed so that the injected gas has the properties required by the recovery system. Often

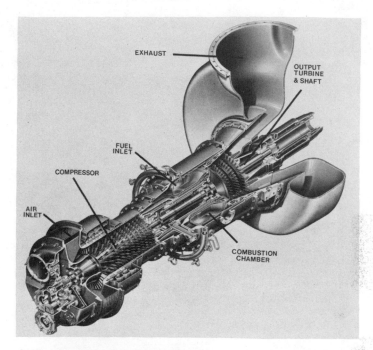

Figure 3-13. Gas turbine engines are used to drive large loads efficiently. (Courtesy of Solar Turbines Incorporated).

enhanced recovery systems require that as the gas mixture enters the reservoir, the gas must be at its critical point—the specific pressure and temperature where both liquid and gas phases can exist simultaneously. The critical pressure and temperature of a gas mixture depends on the components blended in the mixture, and blending, compression, and overall operation must be carefully controlled to maintain the precise conditions for a recovery system to function properly.

Storage Systems

Gas used for injection sometimes must be stored on-site for some time. This is particularly true of gases such as carbon dioxide which are usually trucked or piped in non-continuous shipments. Gas can be stored in high-pressure steel tanks such as those shown in Figure 3-19.

Gases are also stored in underground caverns. These are located in shallow salt formations and are formed by drilling wells into them. Water is first pumped into the formations to dissolve the salt and then pumped out to remove the salt and leave a cavity. The well is cased and uses a wellhead and appropriate valves as shown in Figure 3-20. This arrangement is called a storage cavity or a storage well.

Figure 3-14. A gas turbine engine is enclosed in a housing to prevent noise emission and to provide environmental protection for the equipment.

Gas is not stored in storage wells for great lengths of time. Although salt formations are not usually porous or permeable, gas can leak out of the salt formation if given the time to do so. Also, gas is not stored at high pressure to avoid fracturing the formation and losing the gas. The cavity of a storage well can be quite large; its size depends on the amount of salt removed. A storage cavity may be enlarged by pumping fresh water into the cavity and then back out. Therefore, storage wells are usually very large, inexpensive, low-pressure storage tanks that require no surface space.

Gas Sources

When gases are to be injected for secondary or enhanced recovery, a long-term, reliable, and prolific source for the gases must be secured. Natural gas for use in injection is not usually a problem because it can be taken from gas production facilities serving the lease. However, when the hydrocarbon gases to be used for blending an injection stream must be pure methane, ethane, or propane, the gases must be obtained from a gas processing plant that has the facilities for separating these gases. Because of

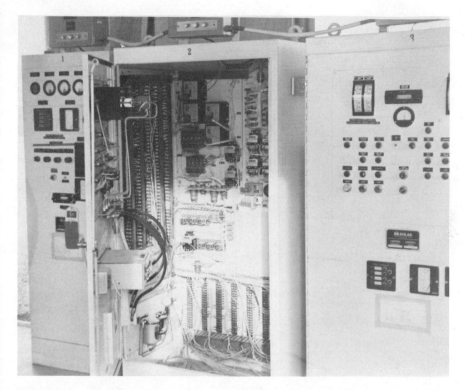

Figure 3-15. A sophisticated control panel operates and protects a gas turbine engine.

the difficulty and expense of using processed gases, most gas injection projects utilize unprocessed natural gas or non-petroleum gases.

Carbon dioxide is commonly employed in enhanced recovery systems using gas injection. It is (for the most part) inert to the materials used in the system and poses no danger to life under normal conditions. Also, since carbon dioxide is not a fuel, there is no reason to justify its use for recovery rather than for fuel.

Large carbon dioxide reservoirs have been located in the southwestern United States. These are capable of supplying the needs of petroleum recovery in this part of the country for many years, but the carbon dioxide is located far from petroleum leases. Carbon dioxide must be compressed and transported long distances before it can be used, and this transportation adds to the cost of the gas. Some producers have been able to lower costs by pooling transportation costs and building pipeline and compression facilities to satisfy all their needs.

Some producers have also developed sources of carbon dioxide by tapping the exhaust stacks of facilities burning large quantities of fuels such as oil,

Text continued on page 60

Figure 3-16. Several gas turbine engines are usually used together in very large water injection stations.

Figure 3-17. The exhaust heat from gas turbine engines may be recovered with heat exchangers.

Figure 3-18. Waste heat recovery from several gas turbine engines may provide heat for treating in a central battery.

Figure 3-19. Gas may be stored at high pressures with steel tanks.

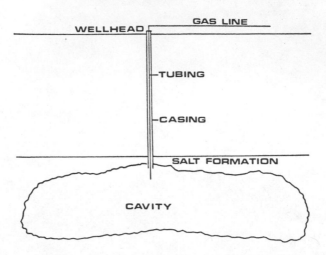

Figure 3-20. Large quantities of gas may be stored at low pressures in storage cavities.

gas, and coal. Power plants, metal smelting plants, and some petrochemical plants are good examples. Unfortunately, these sources do not always provide the large supplies needed. The location of carbon dioxide sources is presently occupying the research facilities of many companies.

Finding other usable gases has been difficult also. Hydrogen sulfide was originally used as one of the gases in one enhanced recovery method. Hydrogen sulfide is prevalent in many petroleum reservoirs and poses a problem in gas processing because it must be removed before natural gas can be used as a domestic fuel. Although hydrogen sulfide can be removed from natural gas, it is difficult to obtain enough of it from most gas plants to satisfy the needs of even modest enhanced recovery systems.

Nitrogen, which is being used as an injection gas to recover condensate from some reservoirs, is taken directly from the atmosphere. Air is compressed, cooled, then processed through a fractionation (a form of distillation) column which separates nitrogen from the other gases composing air. Obviously, there is an ample supply of nitrogen, but the plants required for nitrogen recovery are expensive.

Instrumentation and Control of Gas Injection Systems

For an enhanced recovery mechanism using gas injection to work, a complex blend of several gases must be created. For the critical point of this mixture to match reservoir characteristics and for the gas to sweep the formations effectively, the proportions of the gases must be carefully controlled to exacting standards. Specially designed meters, control devices,

Figure 3-21. The control room of an enhanced recovery system contains electric motor starters and electronic and other control equipment.

and control valves are used to route several gases into a common blending line where they mix to form the correct blend. Figure 3-21 is a photograph of the control room of a small enhanced recovery project. The instruments and control devices are required to analyze the constituent gas streams and to blend them to the correct mixture on a continuous basis.

In the early research into enhanced recovery using gas injection, an interesting problem developed. Once the gases had been mixed and compressed to the required pressure, it was difficult to find a method of measuring the volume of the gas. The injection gas, at its critical point, did not behave as conventional gases. Conventional gas meters simply could not measure injection volume accurately, and it became necessary to develop new methods of metering and controlling gas streams in most gas injection projects.

Orifice metering remains the most reliable method of measuring some gas injection streams. Gas turbine meters are often used accurately on other gas systems. With the problem of critical temperatures and pressures, some gas meters must be connected to computer systems which calculate rather than directly measure volume.

Gas chromatographs (instruments which analyze gas for its components by determining the patterns in which they separate while passing through a series of tubes) are used to determine the makeup of gas streams. This information, along with rates, temperatures, and pressures, can be used to describe precisely the gas to be injected.

Gas Compressors

For the most part, compressors used for gas injection are identical to those used for transportation of gas. The control systems operating these compressors are sometimes more complex than transportation machines because discharge pressures and temperatures are usually critical. Also, the speed and discharge rate from an injection compressor is often controlled from a remote point and requires automatic control devices of some sort.

Injection compressors used in enhanced recovery sometimes have a unique problem not encountered with transportation compressors. The injection pressure and temperature in an enhanced recovery system are sometimes such that both gas and liquid can exist in the injection stream. Because liquid is incompressible, its presence in a positive displacement compressor can cause serious damage to the equipment. In most enhanced recovery systems the compressor's discharge pressure and temperature are such that only gas is actually placed in the discharge header. The gas is expected to cool to its critical temperature before reaching the wells. However, some compressors actually discharge under conditions when liquid could exist in the cylinders. Special clearance cylinders and pistons are required in such cases so that the compressors can be operated without fear of damaging the equipment.

Injection Distribution Systems

Once fluid has been pressured by pumps or compressors, it must be delivered to wells where it can be injected. The piping system providing a flow path to wells is the injection distribution system. A distribution system must be designed to allow fluid to flow to individual wells with no pressure loss due to friction and minimum pipe size to keep installation costs reasonable. The pipe used for injection distribution systems must be chosen to withstand the discharge pressure from the injection station. Also, the pipe must be made of materials that can be buried and left in the earth for long periods of time or covered with materials that protect the pipe from corrosion and possible mechanical damage.

Almost all injection distribution systems are buried instead of laid on the earth's surface. For one thing, the pressures are usually so high that any damage could cause a rupture that could send vast volumes of injection fluid

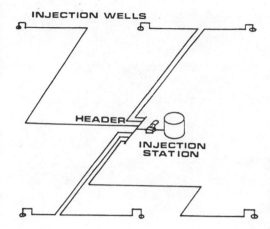

Figure 3-22. All injection lines come from a common control manifold in a radial injection system.

into the atmosphere and onto the land as well as pose a danger to anyone nearby. Also, most such systems are buried to take advantage of the almost constant temperature a few feet beneath the surface. The temperature 3-5 feet below the surface is almost constant throughout a year.

There are several types of pipe arrangements for injection distribution systems. Each type has advantages and disadvantages, both in the long-term operation and the initial installation of the system.

Radial Injection Systems. A radial distribution system is characterized by the line to every individual well going directly to the discharge header of the injection station as shown in Figure 3-22. One advantage of this system is if the injection line to one well requires service, that line can be blocked and serviced without affecting the operation of other wells. Also, wells may be added as needed. Another advantage is that since all injection lines come to a central point, all controls and valves can be placed at a single injection manifold (Figure 3-23), simplifying operation and control of the wells. The major disadvantage is in the cost of literally miles of expensive pipe going to each wellhead. This cost factor can often be so important that it overshadows the advantages.

Axial (Trunkline) Distribution Systems. Another method of distributing fluid to be injected is to use an axial or trunkline system shown in Figure 3-24. The trunkline simply extends the discharge header far from the compressor or pump station, and individual injection lines branch from the central line. The principal advantage of an axial system is its cost: With only one large line instead of many small lines leaving a station, less pipe is required, and the installation cost is far less than that of a radial system. There are several disadvantages. One is if a leak or other maintenance

Figure 3-23. The central manifold of a radial injection system contains control valves and meters for all wells.

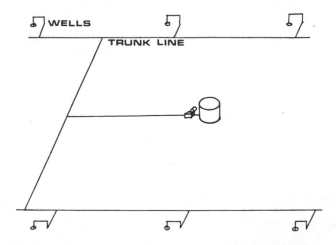

Figure 3-24. Injection lines are supplied from trunklines in an axial injection system.

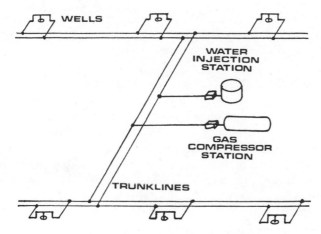

Figure 3-25. Many enhanced recovery injection systems use trunkline injection systems to distribute injection fluids.

problem occurs in the trunkline, the entire system must be taken out of service until repairs can be made. Also, to take advantage of the cost savings of an axial system, the trunkline is seldom sized much larger than needed for the immediate application, and this size restriction can cause pressure loss due to friction. Trunklines can and should be sized larger than actually needed (sometimes as much as twice as large as required), but this lessens the cost advantage of an axial system.

Combination Distribution Systems. Distribution systems can be used which combine the advantages of both radial and axial systems. Individual injection lines are supplied from headers, but the headers are supplied by trunklines extending back to the pump or compression facilities. Figure 3-25 diagrams the distribution system for an enhanced recovery system which uses high-pressure gas but only moderate pressure water. Since water and gas are not injected simultaneously, it is not necessary to provide high-pressure piping for the water.

Types of Pipe Used in Distribution Systems

The pipe used for distribution systems is similar to that used for other applications, but several considerations must be made when the material for a system is specified. First, the pipe must be capable of withstanding the maximum station pressure for extended periods of time. Second, the pipe must be able to transport the fluid with essentially no pressure loss due to friction. Finally, the pipe must be able to withstand the attack of sometimes extremely corrosive fluids for periods of more than 20 years. Several materials are available to meet these criteria.

Plastic-Coated Steel Pipe. The common material for injection systems is steel, but this material is susceptible to corrosion. A layer of plastic is applied to the inside surface of the pipe (sometimes to the outside surface also) to keep fluids from contacting the steel directly. With the proper yield strength of steel, injection lines can be made to withstand any pressure. This method is effective in many applications if the plastic-coating material is thick enough, but there are almost always imperfections in the coating (particularly at couplings) which allow corrosive fluids to atack the steel.

Cement-Lined Steel Pipes. One of the first methods of protecting steel pipe from corrosion was to apply a 0.5-1.0-inch thick layer of cement to the inside surface of the pipe. This method is adequate in some applications but has been largely replaced by the use of plastics. The cement sheath is very fragile and can crack with a slight jar during handling. Once the sheath has been cracked, corrosive fluids can contact and attack the steel.

Fiberglass Pipe. In many applications fiberglass has replaced steel as the predominant material in injection systems. This pipe is available at a pressure rating up to 1500 psi, which makes it acceptable for most water injection systems and some gas systems. Steel must be used for pressures above this level. The principal advantage of fiberglass is its ability to transport virtually any corrosive fluid without degradation. The principal disadvantage is that fiberglass pipe usually costs slightly more than comparably sized plastic-coated steel pipe, but this cost difference is often offset by the long lifetime of fiberglass pipe.

Asbestos Pipe. When large-diameter, low-pressure pipe is needed in an injection system, asbestos (transite) pipe may be used. The internal pressure is not usually allowed to exceed about 200 psi. Also, this pipe is susceptible to mechanical damage. Although this pipe can be used in some applications, it is not designed for most injection systems.

Pressure Rating of Injection System Piping. In most injection pump compressor stations the discharge pressure rating of the equipment is higher than that needed at individual wellheads. In most cases the pressure is reduced by valves located at the wellheads of the injection wells, but sometimes pressure is reduced at the station and delivered to the wells at low pressure. In either case the distribution piping must be designed for the maximum discharge pressure so that it will not rupture if there is a malfunction in the pressure-regulating valves.

Figure 3-26. An injection header allows all controls and meters to be placed at a central location.

Injection Headers

In many injection systems the controls and meters for individual wells are placed at a central point to facilitate operations and data-gathering work. Injection headers are used to accomplish this function. In most radial systems the header is located at the compressor or pump station. In some combination systems several headers may be located throughout a lease. Figure 3-26 shows an injection header located at the injection station. Control valves, meters, and some ancillary equipment are located on the header. The configuration of the wellheads is pictured in Figure 3-27.

Chemical and Polymer Injection

In some enhanced recovery systems special mixtures of chemicals and polymers are required. These chemicals are water soluble and are mixed with the water to be injected. Blending, metering, and controlling this process is often critical. The chemicals used have extremely long molecules which are necessary for the recovery mechanism. If the polymers are blended with water under turbulent conditions, these molecules start breaking apart, and the recovery mechanism's effectiveness is reduced. Even some liquid meters cause enough agitation to break apart polymers. The polymers are routed through special blenders which assure complete and smooth mixtures. Fluid mixtures are metered after mixing.

This blending is usually done at a blending station which may be located either near the pump station or remote to it. Figure 3-28 shows a blending

Figure 3-27. An injection wellhead has only blocking valves when controls are located at a header.

Figure 3-28. A blending station is used with some of the enhanced recovery injection methods.

station in a major enhanced recovery system. The tanks and bulk bins are used to store the dry chemicals and clean water.

Control Valves

Pressured fluid sent to a well for injection usually arrives at a much higher pressure than is required for the particular well. A control valve is installed

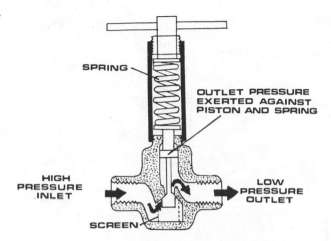

Figure 3-29. A pressure control valve maintains a constant downstream pressure regardless of throughput.

for each well to reduce this pressure under controlled conditions. The valve must deliver fluid to the wellhead at a constant pressure or rate so that when fluid reaches the bottom of the well, the total pressure (including hydrostatic head) is enough to push fluid into the reservoir without fracturing it.

Pressure Control Valves

Figure 3-29 shows one type of pressure control valve used to control fluid flow. Downstream pressure against a small plunger opposed by the manually set tension of a spring operates a needle valve arrangement. When the spring's tension has been set, the valve will hold constant downstream pressure regardless of upstream pressure, provided the upstream pressure is higher than the downstream pressure. Figure 3-30 shows another control valve which utilizes downstream pressure exerted against a diaphragm and opposed by spring tension to control fluid flow.

Rate Control Valves

Figure 3-31 shows a valve used for maintaining a constant flow rate through the valve. Basically, the valve uses internal channels to sense the pressure drop and holds this difference constant by exerting pressure against a plunger opposed by spring tension. The pressure difference across the valve should be constant across the sections of the valve used for control if the rate is constant. This particular valve can be converted from rate to pressure control by adding a diaphragm operator.

Figure 3-30. A pressure control valve may be a diaphragm-operated instead of spring-operated unit.

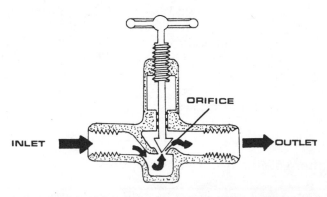

Figure 3-31. The size of the orifice in a constant rate control valve determines the controlled rate.

Figure 3-32. Control of injection wells may be accomplished using electronic instruments and control valves.

Electronically Controlled Valves

In the last few years improvements in the area of field electronics have made it possible to provide control for injection wells which can be either pressure or rate controlled, or a combination of both, through the use of electronics. Figure 3-32 shows a control valve arrangement, including the valve, rate and pressure detectors, and the electronic circuits necessary to control the valve. The particular arrangement controls the valve to hold a constant rate, unless the pressure exceeds a certain set point, at which time the controller starts holding constant pressure on the wellhead.

Cavitation in Injection Well Control Valves

Cavitation is the process when the pressure difference across a valve is sufficient for free gas (water vapor or the gaseous form of the injection fluid) to be liberated in the valve. This free gas interferes with the operation of valves which are intended to operate with only liquid, and it can also cause erosion problems in the operating parts of the valve. Cavitation can sometimes be the most serious operational problem in injection control

Figure 3-33. Several valves are usually used on injection well-heads to provide ease of maintenance and operation.

valves, and care must be given to selecting valves and operating conditions to avoid such problems. Many control valves are available which have special cages and baffles to reduce the likelihood of cavitation problems.

Other Valves in Injection Wellhead Arrangements

Several other valves besides the control valve are required for most injection wells. Figure 3-33 shows a wellhead with several valves. The valve on top of the injection tubing is a full-opening gate valve which, when properly sealed from the atmosphere, allows the passage of subsurface logging and service tools into the tubing string. There are also isolation valves which can be closed to separate the wellhead control equipment from the high pressure of both the injection system and the tubing string. Valves are sometimes installed so that future piping can be connected without taking a well out of service.

Data Collection

In most contemporary injection systems it is critical that the injection rate and pressure be monitored on a regular basis. The injection rate and

Figure 3-34. Pressure gauges monitor wellhead pressure.

pressure are recorded by operators at least weekly and usually daily. Thus, injection wells must be equipped to monitor both rate and pressure.

Most injection wells are equipped with pressure gauges like the one shown in Figure 3-34. Gauges are provided to measure downstream or wellhead pressure. They may also be provided to measure injection system pressure upstream of the wellhead control valves. Finally, a pressure gauge is usually provided to monitor the pressure of the tubing-casing annulus to assure that the subsurface packer is still functioning properly.

To measure the rate at which fluid is injected into a well, a fluid meter of some kind must be used. Liquid meters may be positive displacement meters or paddle-type meters (Figure 3-35). Turbine meters may also be used, since they are durable and accurate for such service. Figure 3-36 shows a turbine meter and a portable rate meter that can measure the instantaneous rate. Figure 3-37 shows another turbine meter with a permanently mounted electronic display unit which shows both instantaneous rate and total volume metered.

When gas is injected into a well, the gas must be measured by different methods than those used to measure liquid. The most common method of

Figure 3-35. Paddle meters monitor injection volumes in wells.

measuring injected gas is with orifice meters as illustrated in Figure 3-38. This particular meter has a built-in integrator which converts the pressure and differential pressure data into volume increments to drive a counter. Another method of measuring gas is to use gas turbine meters (Figure 3-39). Of course, the positive displacement meters discussed in Volume 2 may also be used.

Ancillary Equipment for Injection Wells

Most injection wells require other equipment besides control valves, meters, and gauges. Most meters, some control valves, and a few pressure gauges cannot function if solid material such as sand is allowed to pass through them. Filters or strainers are used to prevent the passage of such material. Figures 3-40 and 3-41 show in-line filters and strainers, respectively, in water injection service.

When water—even salty water—is injected into a well, there is the possibility that the water can freeze in the wellhead during cold weather. Gas flowing through a control valve cools and can also freeze in cold weather. Thus, most injection wells are provided with some means of freeze

Figure 3-36. Liquid turbine meters measure liquid injection rates.

protection. This protection may be as simple as wrapping the pipes with heat-tracing tape (high-resistance wire to which 120-volt power is applied to generate heat) and insulating the pipes. In some climates specially constructed metal enclosures (freeze boxes) are installed over the wellhead and internally heated.

Methods of Controlling Injection Wells

When fluid is injected into a formation, fluid may be allowed to flow so rapidly that it fractures the rock in much the same way that hydraulic stimulation procedures do. If a fracture is allowed to begin and propagate through the reservoir, the efficiency of the recovery mechanism may be diminished because some of the formation may be by-passed by injection fluids as they flow through the fracture(s). Ultimately, every injection control method attempts to place as much fluid in the reservoir as possible without fracturing the rock.

Essentially, there are two ways of controlling the rate and pressure of injection wells. One is to maintain a constant injection pressure at the wellhead and allow the rate to fluctuate, and the other is to hold the rate constant and allow the pressure to fluctuate. Many operators absolutely

Figure 3-37. Turbine meters and permanent detectors measure both rates and volumes of liquids being moved.

insist on one or the other of the two methods; therefore, some explanation of the difference between the two is necessary.

Constant Pressure Control

Constant pressure control amounts to holding the surface injection pressure at a constant level for long periods of time. If the pressure was not enough to fracture the formation originally, the thinking is that it will not fracture the formation later in the life of the well.

For fluid to be injected into a formation, the pressure in the wellbore must exceed the pressure in the heart of the reservoir. As fluid is injected into and accumulates in a reservoir, the reservoir pressure increases. If the wellbore pressure remains constant and the reservoir pressure increases, the net difference between the two must decrease, and the flow rate—whose value is directly related to the difference in pressure—must decrease. A characteristic of constant pressure control is that the flow rate decreases with time.

Figure 3-38. Orifice meters measure gas flow in injection systems.

Figure 3-39. Gas turbine meters measure gas rate and volume in gas injection systems.

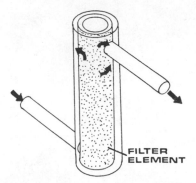

FILTER
ELEMENT

Figure 3-40. A filter is placed in an injection line to remove fine solid materials.

STRAINER

Figure 3-41. An in-line strainer removes larger solid particles.

If the objective of a recovery system is to put as much fluid into a reservoir in as short a period of time as possible without damaging the reservoir, constant pressure control is not the most efficient method available. However, there can be no doubt that a formation will not be fractured in time by constant pressure control because the flow rate—which actually controls the fracturing process—decreases with time.

Constant Rate Control

As fluid is injected into a reservoir, the reservoir pressure increases. Therefore, to maintain a constant injection rate, the wellbore injection pressure must increase to keep the difference between the two pressures constant. A characteristic of constant rate control is that the injection pressure increases with time.

When a well is first placed in injection service, the bottom hole pressure is low because the well was probably a production well beforehand. When fluid first begins to enter the reservoir, little wellbore pressure—less than the hydrostatic head of tubing fluid—is required. The well has no surface injection pressure, and the only way to control the well's operation is to maintain the rate at some fairly low level. As the well stays on constant rate control, the injection pressure increases to the point that some surface injection pressure is required.

Some operators claim that if the surface injection pressure increases to high levels, the injection well will fracture the formation. However, if the rate was not high enough to fracture the rock originally, it will not fracture later as the reservoir pressure increases. In an injection application, as fluid is injected into the reservoir, the saturation of that fluid must increase, making the relative permeability to that fluid increase. In other words, as fluid is injected, the resistance to fluid flow decreases, and the fluid is less likely to fracture the formation.

From the standpoint of putting as much fluid in a reservoir as possible in the shortest possible time, constant rate control is much more efficient than constant pressure control. Provided the injection rate is selected at a safe level, there is no reason that the formation should ever be fractured regardless of the injection pressure finally achieved.

Wellhead Equipment Arrangement

From the preceding discussions, several pieces of equipment are required to work an injection well. When the equipment is located at the wellhead, some logical arrangement must be used to be sure that all equipment works together properly.

Figure 3-42. Several valves, controls, and meters are usually required on injection wells.

Water Injection Wells. Figure 3-42 shows an arrangement of wellhead injection equipment which allows all equipment to function properly. A single valve is used to isolate all equipment from the injection system. Downstream of this valve are the control valve, pressure gauges, and meters. The tubing is equipped with a full-opening gate valve to provide a means of lowering tools into the tubing. This valve is also used to isolate tubing pressure from the control and metering equipment for servicing.

Gas Injection Wells Figure 3-43 shows a common arrangement of equipment for a gas injection well. The arrangement is much the same as for a water injection well, except that the equipment is designed for higher pressure. The orifice meter is separated from valves and fittings by long pieces of straight pipe so that the meter functions properly.

Enhanced Recovery Injection Wells. Most enhanced recovery systems require alternate injection of water and other fluids (gases or polymers). These wells may be configured as shown in Figure 3-44 to provide alternate flow paths.

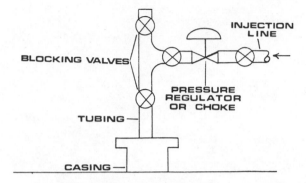

Figure 3-43. A gas injection wellhead requires control valves and blocking valves for proper operation and maintenance.

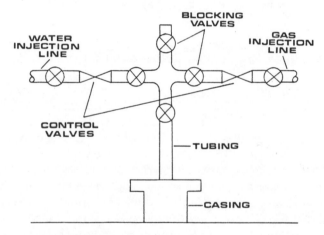

Figure 3-44. Enhanced recovery injection wellheads usually require duplicate control and block valve arrangements to handle multiple injection fluids.

Subsurface Equipment

The subsurface equipment required for most injection wells is the same as that discussed for saltwater disposal wells. A very simple arrangement of a tubing string and a packer are all that are required. The packer is necessary to isolate high injection pressures from the casing, which must stay in the wellbore for many years without replacement. The annulus of most injection wells is usually filled with a mixture of corrosion inhibitors to prevent corrosive attack on the tubing, casing, and packer.

Figure 3-45. Central battery functions and injection facilities are incorporated into a single facility on most large leases.

Combined Production, Disposal, and Injection Facilities

When production treating, water disposal, and injection systems are installed for major production leases, many operators prefer to place all equipment in a single location so that various systems that must interact with each other are physically close to their counterparts. Locating major systems close to other systems also allows for easier operation of the entire complex (as well as providing a means of sharing certain utilities such as electrical power, instrument and control rooms, and waste heat recovery systems).

Figure 3-45 is a central lease facility in which crude oil is treated, water is accumulated for disposal, and other water is conditioned and pumped to wells for injection. The control room houses equipment used in all systems, and utilities such as fresh-water supplies, electricity, instrument air for pneumatically operated devices, and fuel gas are provided at a central point. Many companies build and operate their own gas processing plants, and it makes sense to locate these plants close to the source of the natural gas—the central processing facility.

Control of Complex Facilities

In the first three chapters of this volume we have seen processing and other systems that are much more complex than the simple systems discussed in Volume 2. The evolution of complex systems has been necessitated by the increasing expense of constructing and operating processing, transportation, storage, and operating facilities and by increasing prices of crude oil and natural gas which demand efficiencies unprecedented in the petroleum industry. Control and operation of these systems is possible using the simple valves and other devices, but the manpower requirements of such operations is not feasible. Chapter 4 discusses instrument and control systems capable of handling the complex systems discussed in this volume. Also discussed are interpretations of some of the data related to equipment in this volume as well as in Volume 2.

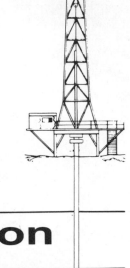

Chapter 4

Instrumentation and Control

As the complexity of petroleum production facilities increases, the methods of controlling these facilities must become more sophisticated. Although manual control devices (some valves and level gauges) are used in even the most complex control systems, a number of devices not previously discussed are used in most modern systems. This chapter discusses several control and measurement devices that have been previously mentioned as well as a number of devices not yet introduced. Most instrumentation and control equipment to be discussed falls into one of three categories: (1) pneumatic, (2) electrical, and (3) mechanical. However, some devices are combinations of one or more of these. Some control and instrument devices are manually operated (requiring human intervention), while others are automatic (capable of unattended operation).

Mechanical Control and Instrumentation Systems

Mechanical instrument and control devices are usually distinguished by the absence of pneumatic and electric components. Such equipment operates using springs, levers, baffles and flow channels, and hand wheels. By this definition, a treater is an automatic, mechanically controlled process vessel because its operation is controlled by baffles and channels inside it. Most manual control devices are mechanical in nature.

Pneumatic Instrumentation and Control Equipment

Pneumatic control equipment operates when gas under pressure pushes against a surface of some kind (such as a diaphragm). This surface then pushes against the parts to be moved. One reason that pneumatic equipment is so often used in petroleum production operations is that a convenient source of pressured gas (produced gas) is nearly always available. Natural gas is not always a good pneumatic supply because it often contains corrosive gases (such as hydrogen sulfide) or by-products of iron oxidation which can plug small orifices in the equipment.

Air is an excellent pneumatic supply gas because it seldom contains corrosive components or materials that can plug equipment, but air must first be compressed for use. Because air usually contains water vapor, compressed air should be processed through a dehydrator before it is used in pneumatic equipment to prevent condensation and freezing from interfering with the equipment's operation.

Instrument Air Compressors

When used to operate pneumatic devices, air is called instrument air. Instrument air is usually supplied by fairly small, reciprocating (positive displacement) or rotary (centrifugal) compressors. Air is taken from the atmosphere, compressed, dried as needed, and sent to a volume tank which simply holds enough air to allow the pneumatic control system to have an uninterrupted supply. The compressor is usually driven by an electric motor which turns on and off depending on the pressure in the tank. The pressure in the tank is held high enough that all devices have an ample supply of instrument air. Dehydrators are usually integral parts of the instrument air compressor system. Figure 4-1 shows a small compressor supplying instrument air to a small process facility, while Figure 4-2 shows a large, skid-mounted compressor used for much larger facilities.

Electric Instrumentation and Control

In the last few years a number of electric instrument and control devices have been developed for use in petroleum operations. Previously, electric equipment was considered too expensive, too delicate, and too complicated for such use. However, with the increase in the level of technology in both the petroleum production industry and in the electric manufacturing industry, this is no longer the case. Electric control and instrument equipment usually has the advantage that it requires no special compression devices to supply energy to operate the equipment. Rather, a simple 120-volt supply, such as is used to operate lights, is all that is required.

Figure 4-1. Small, instrument-air compressors are used when pneumatic loads are light.

Figure 4-2. Larger compressors are required when there are many pneumatic instruments and controllers.

Electric equipment is no more dangerous or complex to trained personnel than is pneumatic or mechanical equipment.

Control Valves and Valve Operators

The major types of valves that are used in petroleum production systems were discussed in volume 2. Most of the valves used in production facilities are like those shown on the wellhead in Figure 4-3. That is, most are small, hand-wheel or handle-operated isolating and maintenance valves that are seldom adjusted. The major criteria for selecting manual over automatic operated valves is the frequency, accuracy, and repeatability of adjustment.

When valves must be adjusted frequently or continuously or when their control is critical to a process, automatic valve operators are often selected. Mechanical, pneumatic, and electric valve operators are available for all valves normally used in production activities. Selection of the type of operator is made based on operating conditions.

Mechanical Valve Operators

Manual valve operators are hand wheels or handles attached to the valve stem. Such operators can be used to move linear and rotary valves. Some valves are equipped with manual operators that can be replaced with other types.

Most automatic mechanical operators consist of lever systems attached to the valve stem. Figure 4-4 shows a wafer valve used as a dump valve which is operated by a float through a linkage. As the float rises, the valve opens; as the float falls, the valve closes. This is one common method of controlling the level of liquid in a separator.

Figure 4-5 shows a mechanically operated valve used to control the flow into an injection well. This operator functions by forcing water pressure to push against a spring whose tension is adjusted to cause the desired downstream pressure.

Pneumatic Valve Operators

Pneumatic operators can be connected so that a person must be present for the operator to function, or they can be configured to function without human intervention. Pressure regulators and back-pressure regulators are automatic, pneumatically operated valves. In some smaller regulators and pilot valves the operator is an integral part of the valve body, while in larger applications the operator is a separate part attached to the valve.

Figure 4-6 shows a diaphragm operator on a manifold valve. This valve is used for well test headers in test facilities. Figure 4-7 shows a large

Figure 4-3. Hand-operated valves (like those on wellheads) are the principal valves for control throughout petroleum production operations.

Figure 4-4. The dump valve of a separator is a wafer valve operated by a float.

Figure 4-5. A mechanically operated valve controls the flow of liquid into an injection well.

Figure 4-6. Diaphragm operators can operate valves on a welltest header.

Figure 4-7. Large automatic or emergency valves usually require large pneumatic operators.

pneumatic actuator on a gate valve controlling the water supply in a large injection system.

Pneumatic valve operators are particularly effective when rapid valve movement is needed. Figure 4-8 shows a separator dump valve arrangement using a "snap-acting" valve (a very fast valve). The float valve operates a small pilot valve—just a pneumatic switch—which supplies gas to the diaphragm-operated dump valve. As soon as the float reaches a predetermined position, the pilot switches and forces the diaphragm valve to move. The mechanical arrangement in Figure 4-4 causes the dump valve to open and close fairly slowly. If the separator were to drain rapidly, gas could enter the liquid system unless the dump valve switches rapidly.

Electric Valve Operators

Electric valve actuators consist of electric motors which are attached to the valve shaft. Figure 4-9 shows the electric actuator for a rotary valve such as a ball or wafer valve. The actuator consists of an electric motor with two

Figure 4-8. Pneumatic operators are used when high speeds are necessary as in a snap-acting dump valve.

Figure 4-9. Electric actuators are well suited for electrically operated systems such as welltest headers.

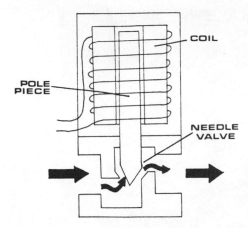

Figure 4-10. A solenoid valve uses electromagnetic action on a small needle valve to route gas through the solenoid.

sets of windings (so the motor can turn in either direction), and a set of limit switches which stop the motor when the valve shaft reaches the desired position.

Most electric actuators are slow in comparison to pneumatic operators and may take several seconds to move a valve from one position to another. For some applications, such as separator dump valves, this speed is much too slow, and pneumatic operators are preferred.

Solenoid Valves. A solenoid valve is an electric actuator which consists of a coil, a pole piece, and a small needle valve as shown in Figure 4-10. When electric current flows through the coil, it becomes an electromagnet and moves the pole piece, which in turn moves the needle valve to switch supply gas to pneumatic devices. Solenoids operate very quickly (usually faster than pneumatic devices) and are often employed to act as pilot valves for pneumatic operators. Figure 4-11 shows a solenoid being used to switch a diaphragm operator which operates a valve.

Level Detectors and Control Devices

Many processes in petroleum production depend on the level of liquid in vessels. There are a number of devices used to detect and control liquid levels. Most level detectors and control devices are pneumatic or mechanical in nature.

Level Detectors

Floats. Liquid floats are the most common method of detecting liquid levels. Figure 4-12 shows a float which can be mounted in virtually any vessel to detect liquid level. This particular float is equipped with a small gas switch

Figure 4-11. A solenoid is usually used to switch instrument air to pneumatic operators.

which can be used to supply a pneumatic device. The gas switch can be replaced with an electric switch so the float can drive electric devices. Figure 4-13 shows a similar float equipped with weights. This float is used to detect the interface between oil and water. It is weighted to float on water but sink in oil.

Hydrostatic Head Detectors. Many control applications require that it be known not only when liquid reaches a certain level, but also how deep the liquid stands. Figure 4-14 shows a device used to indicate the standing level and to operate a switch when the liquid reaches a certain point. The hydrostatic head of liquid exerts a force against the diaphragm which operates a pointer and a switch mechanism. The adjustment wheels allow the switch to operate at varying depths. This device can be used to turn on a transfer pump when liquid reaches a predetermined depth in a tank and turn off the pump when the level reaches another depth.

Figure 4-12. Float valves are the most common method of maintaining the level in a vessle.

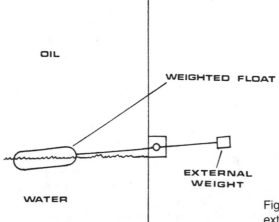

Figure 4-13. A weighted float and external weights are balanced so the float sinks in oil and rests on the water at the interface.

Figure 4-14. Hydrostatic level detectors indicate liquid level as well as providing for level control.

Electronic Level Detectors (Resistive). Figure 4-15 is a sketch of the operation of an electric level detector which can detect the level of water in a vessel. When water covers both probes, they become "shorted" together (there is no resistance between them, and they act as a closed switch) and can be used to switch electric control devices. This method does not work with oil levels because oil does not conduct electricity and cannot short the probes.

Electronic Level Detectors (Capacitive). Figure 4-16 shows a somewhat more complex electric level detector which operates on the principle of capacitance (the ability of a substance to store an electric charge). When liquid covers the probes, the capacitance between them changes, an electronic circuit detects the change, and an internal electric switch operates to drive electric control devices. Capacitive level detectors tell not only when liquid reaches the probes, but how deep liquid stands over them. Thus, these detectors can be used to determine the depth of liquid in a vessel.

Electronic Level Detectors (Differential Pressure). Figure 4-17 shows another electric level detector which functions by measuring the difference in pressure between the top and bottom of a vessel. Of course, this difference is the hydrostatic head of the liquid. If the density of the fluid in the vessel is known, the exact level can be measured. This device uses

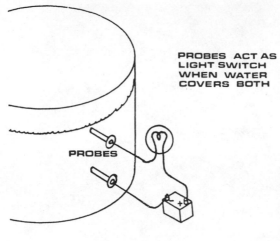

PROBES ACT AS
LIGHT SWITCH
WHEN WATER
COVERS BOTH

PROBES

Figure 4-15. Resistive level switches turn on lights or provide control action when liquid covers both probes.

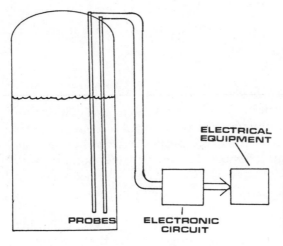

ELECTRICAL
EQUIPMENT

PROBES   ELECTRONIC
CIRCUIT

Figure 4-16. Capacitive level detectors use electronics to sense the capacitance between two probes and determine liquid level.

strain gauges—electronic devices whose resistance changes as they are stretched—mounted on a diaphragm. As the differential pressure changes, the diaphragm stretches and causes the gauges to stretch. An electronic circuit detects the change in resistance and converts it to an electric signal which can switch electrical equipment or transmit data to a remote point.

Sight Glasses. Some manual method of liquid level measurement is usually provided for most vessels. One of the simplest methods is to use sight glasses such as the one shown in Figure 4-18. Liquid in the vessel enters the glass just as inside the vessel. Thus, the internal liquid level is made visible from the outside.

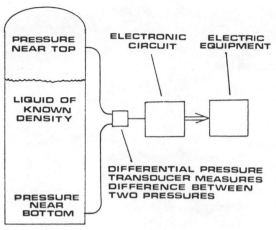

PRESSURE
NEAR TOP

ELECTRONIC
CIRCUIT

ELECTRIC
EQUIPMENT

LIQUID OF
KNOWN
DENSITY

DIFFERENTIAL PRESSURE
TRANSDUCER MEASURES
DIFFERENCE BETWEEN
TWO PRESSURES

PRESSURE
NEAR
BOTTOM

Figure 4-17. Differential pressure level detectors measure the difference in pressure from top to bottom of a known liquid to determine its depth.

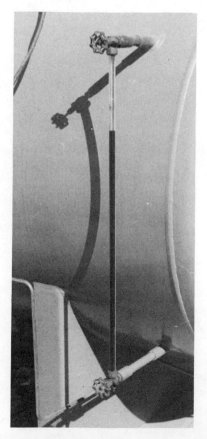

Figure 4-18. Sight glasses allow production personnel to monitor internal liquid levels.

Figure 4-19. A pneumatic liquid level controller uses a float to sense liquid level and then exerts pneumatic control over valves to maintain a constant level.

Liquid Level Controllers

Control devices are available which measure liquid level and control valves or other devices based on this level. For example, a liquid level controller might measure the liquid in a vessel and adjust the position of the valve, allowing fluid into the vessel to control the level. Such control devices are usually continuous devices: if the level changes slightly, the valve is adjusted only slightly; and if the level changes considerably, the valve is adjusted accordingly.

Figure 4-19 shows a pneumatic liquid level controller. A float is attached to the back of the controller and rests in liquid inside a vessel. As the float moves, the flow of instrument air to the control valve is adjusted. Instead of switching instrument air to a control valve, this device throttles air to the valve.

Pressure Measurement

The most common method of measuring pressure in equipment is with pressure gauges. There are a number of types, ranges, and accuracies of pressure gauges.

Figure 4-20. A pressure gauge is the most common method of measuring internal pressure.

Pressure Gauges. Figure 4-20 shows a pressure gauge commonly used for petroleum production operations. Pressure causes a circular tube to expand and move the hand along the scale. Such gauges measure what is called *gauge pressure* (absolute pressure minus atmospheric pressure). A gauge exposed to atmospheric pressure will indicate zero psig. Pressure gauges are available in full-scale ranges from a few ounces per square inch to thousands of psig. As their name implies, oil-filled gauges are filled with a viscous oil which prevents internal damage due to vibration and shock.

Vacuum Gauges. Another type of pressure gauge is the vacuum gauge. This gauge actually measures pressure below atmospheric on a scale. Very accurate vacuum gauges are available, but the range of vacuum gauges does not exceed 14.5 psi, since there is no pressure below zero psia.

Differential Pressure Gauges. Pressure and vacuum gauges measure pressure exerted inside a coil that is closed on the opposite end. Differential pressure gauges operate when two pressures are exerted at the ends of the coil. The coil then responds to the difference in the two pressures and drives an indicator accordingly. Differential pressure gauges (DP or ΔP gauges) then measure the difference in pressure between their two connections.

Pressure Gauge Accuracy. Pressure, vacuum, and differential pressure gauges can be obtained with almost any desired accuracy. For most

Figure 4-21. An ordinary thermometer measures internal
temperature.

applications, such as vessel pressures, gauges that give indications within
5% of true pressure are adequate. Some pressure measurements are more
critical, and readings within 1% of true pressure are needed. Seldom are
more accurate pressure measurements required. In general, the most
accurate gauges are delicate and expensive. Thus, these gauges are not
normally used in routine measurements.

Temperature Measurement

For most production temperature measurements, ordinary thermometers
are used. Figure 4-21 shows an ordinary thermometer. When temperature of
a pressured vessel is being measured with a thermometer, it is not normally
desirable to expose the device directly to the contents of a vessel. In such
cases temperature is measured with a thermowell into which a thermometer
is inserted. Figure 4-22 shows the use of a thermowell to isolate a temperature
detector from vessel fluids.

Temperature Control

When burners are used with fired vessels, the burners are controlled by
devices sensing temperature in the vessel. A thermostat like the one in
Figure 4-23 is often used for burner control. The temperature bulb is a sealed
metal tube connected to the thermostat with a long, hollow capillary tube.
Air in this tube expands with heat and activates miniature valves within the
thermostat to control fuel gas flow and ignition.

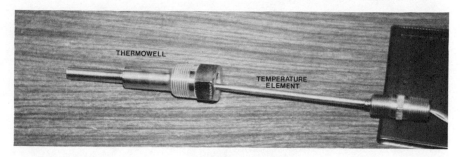

Figure 4-22. A thermowell protects a thermometer or other temperature detector from high pressure and corrosive liquids.

Figure 4-23. A thermostat assembly is often used to control the termperature in a vessel.

Electronic Instrumentation

In the last few years a number of electronic instrument systems have been developed for use in production operations. Electronic instrumentation is well suited for petroleum production operations because it can be made hardy and reliable. Also, this type of instrumentation is quite versatile, and several functions can often be accomplished with one instrument.

Electronic Pressure Measurement

Electronic Pressure Transmitters. There are several ways that pressure and vacuum can be measured using electronic devices. One of the most common ways is to exert pressure against a metal diaphragm to which strain gauges have been bonded. As the diaphragm stretches, the gauges stretch and signal an electronic circuit, which converts the resistance change to a signal that may be used for control or measurement.

Electronic Differential Pressure Transmitters. Like pressure gauges, electronic pressure transmitters use pressure exerted from only one source. Electronic differential pressure transmitters use two pressures on opposite sides of a diaphragm to stretch it and the attached strain gauges. Again, internal electronic circuits condition the signals from these gauges and signal or control other devices.

Electronic Temperature Measurement

Thermocouple Temperature Detectors. Thermocouples are temperature elements consisting of two dissimilar metal strips bonded together. When these strips heat, they generate a slight voltage difference between them. Thermocouples are mounted in sealed steel tubes and inserted in thermowells. Wires from the thermocouple are connected to electronic circuits that detect the voltage difference and display the temperature and signal control devices.

Resistance Temperature Detectors (RTDs). Another type of electronic temperature detector is a resistance temperature detector (RTD). These are small devices in which resistance changes with temperature. RTD's are mounted in tubes in thermowells and connected to electronic circuits which detect the resistance changes and signal display or control equipment.

Chromatographs

In many gas processing and blending systems it is necessary to know the composition of fluids being handled. Chromatographs are instruments

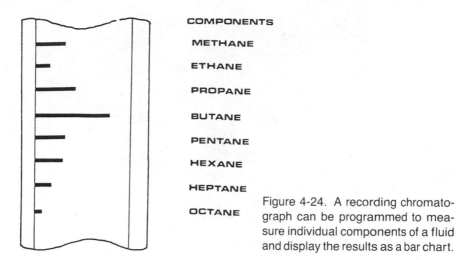

COMPONENTS

METHANE

ETHANE

PROPANE

BUTANE

PENTANE

HEXANE

HEPTANE

OCTANE

Figure 4-24. A recording chromato-graph can be programmed to measure individual components of a fluid and display the results as a bar chart.

which separate gas into constituent parts and measure the proportion of each. The results can be displayed on bar charts (Figure 4-24) or can be communicated to other equipment that analyzes these results. Chromatographs are also available for liquid streams, but they use the same principles as gas chromatographs because they first boil the liquid to vapor before analyzing it. Chromatographs are sometimes necessary to critical processes, but their expense usually prohibits use in ordinary petroleum production equipment.

Atmospheric Emissions Instruments

In populated areas it is necessary to monitor the areas around petroleum production facilities to assure that poisonous or flammable gases are not being released to the atmosphere. Normally, such gases are not released to the atmosphere; a release usually occurs only if a leak develops. In most open areas it is not necessary to monitor for leaks of this sort because the emissions are too small to be of any consequence. However, in populated areas it is necessary to monitor for even the slightest leaks because of the danger to human life.

Hydrogen Sulfide Detectors

Hydrogen sulfide detectors (Figure 4-25) are used to monitor continuously the atmosphere for lethal concentrations of gas. Several are usually positioned around a facility to detect the gas regardless of wind direction. These detectors can be adjusted to trigger an alarm at almost any concentration. They are usually set to alarm at the minimum toxic concentration (about 20 ppm) so that even if it takes a few minutes for

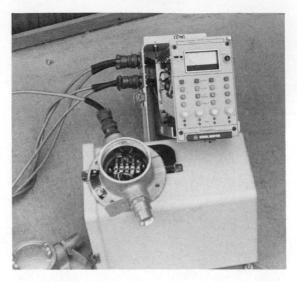

Figure 4-25. Hydrogen sulfide detectors are used for emergency warnings and equipment shutdown. (Courtesy of General Monitors, Inc.)

equipment to be serviced, the gas can dissipate in the atmosphere to less than lethal levels by the time it reaches populated areas.

Portable Detectors

Portable detectors (Figure 4-26) are used to sample suspected hostile atmospheres. These detectors can be used to detect poisonous gases or to detect explosive concentrations of flammable gases such as natural gas.

Vibration Measurement

Some large equipment, such as compressors and large pumps, can be severely damaged by excessive vibration. Vibration switches like the one in Figure 4-27 are used to detect vibration and either turn off motors or indicate excessive vibration.

Load Measurement

Sometimes it is necessary to measure load on structures or other equipment. One particular example is the walking beam of a pumping unit (the load is indicative of the polished rod load). Strain gauges bonded to

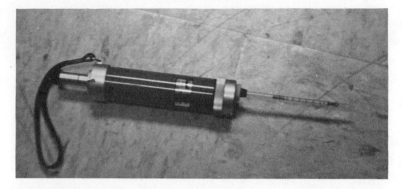

Figure 4-26. Portable detectors are available to check atmospheres for many lethal and explosive gas mixtures.

VIBRATION
SWITCH

Figure 4-27. Vibration switches stop equipment before it can be damaged by excessive vibration.

metal structures stretch and compress with the equipment and can be used to measure the forces being exerted on structural members.

Dynamometer Measurement and Interpretation

Dynamometers (introduced in Volume 2) are instruments used to measure the polished rod load versus the position of the sucker rod lift system. Figure 4-28 is a dynamometer which is placed under special clamps on the carrier bar (Figure 4-29). Dynamometers are used to record what are called *dynamometer cards* or *dynagraphs,* an example of which is shown in

Figure 4-28. The dynamometer is an instrument that measures the performance of beam pumping systems.

DYNAMOMETER
FITS HERE

Figure 4-29. A dynamometer is inserted in the carrier bar of the bridle of a pumping unit.

Figure 4-30. In recent years a number of electronic systems have been developed which can measure the same data as the mechanical dynamometer. Figure 4-31 shows a complete system including a clamp-on strain gauge unit to detect rod load, a position transducer, and an electronic recorder system. Although the recordings are produced from electronic media, such graphs are also called dynagraphs.

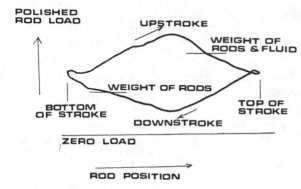

POLISHED
ROD LOAD

UPSTROKE

WEIGHT OF
RODS & FLUID

WEIGHT OF RODS

BOTTOM
OF STROKE

TOP OF
STROKE

DOWNSTROKE

ZERO LOAD

ROD POSITION

Figure 4-30. The dyna-
mometer makes a record-
ing of polished rod versus
polished rod position of a
beam pumping system.

STRAIN GAUGE
TRANSDUCER

POSITION
TRANSDUCER

RECORDER

Figure 4-31. An electronic instrument system is available for measuring the performance of a beam
pumping system. (Courtesy of End
Devices, Inc.)

Dynagraph Interpretation

A dynagraph can be used to determine the subsurface action of the rod string, pump, tubing, and other equipment. The data can also be used to calculate and measure loading of the surface pumping system.

Traditionally, the perfect dynagraph—indicating perfect loading and efficiency in the pumping system—is a parallelogram as mentioned in Volume 2. Realistically, perfect-operation systems cannot be achieved, but many pumping systems may be made very efficient. Figure 4-32 is the dynagraph of a pumping system operating well. During the upstroke, the rods stretch as the dynagraph approaches peak polished rod load. After the peak load, the rods relax somewhat during the remainder of the upstroke. As the downstroke begins, the traveling valve opens, transferring fluid load to the tubing; however, the rods continue to relax in the first part of the downstroke. Late in the downstroke, the rods relax even beyond the point at which they would be if they were hanging motionless in the well.

The dynagraph gives a great deal of information about the pumping system. For example, the area contained within the recording is proportional to the work done by the pumping system to lift fluid. If the area is large, the pump is probably properly loaded and functioning at high efficiency. If the area is fairly small, either the pump or tubing is leaking, and the pump is doing very little useful work; or the well is flowing, and the pump is doing little work. The dynagraph also shows the maximum load on the surface pumping unit and can be used to calculate load and torque data throughout a pumping cycle.

Several factors influence the shape and size of a dynagraph. Pumping speed can affect the shape by establishing harmonic motion in the rod string. Pump depth influences the size of the card because the pump must do more work to lift fluids from great depths. Friction of fluid in the tubing, rods rubbing against the tubing, and mechanical and fluid motion in the pump also cause the dynagraph to take on different shapes. Finally, fluid entry into the pump can cause the pump to take drastically different shapes.

From the standpoint of perfect pumping system operation, all of the factors which affect card shape could be considered to be abnormal. However, most wells have characteristics that cause some unusual shapes of some kind. In fact, the perfect pumping system of Figure 4-32 is more often the exception than the rule. This is not to say that less than perfection should be accepted; however, realistically, some abnormalities must be accepted.

Fluid Pound. One of the most damaging conditions that can occur in a pumping system is fluid pound (mentioned in Volume 2). Fluid pound occurs when the pump does not fill with fluid completely during the upstroke. During the downstroke, the traveling valve stays closed and the

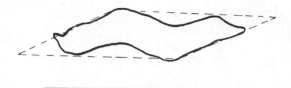

Figure 4-32. The dynagraph of a normally operating pumping system may be shaped roughly like a parallelogram.

polished rod load stays high. When the traveling valve does strike fluid, it opens quickly, and the polished rod drops sharply, causing a jerking effect in the entire system. This condition is illustrated in Figure 4-33. Fluid pound is at its worst when the traveling valve opens near the middle of the downstroke when the rod string is moving at its peak velocity and when the sharp load change places the greatest shock load on the system. Fluid pound occurring early in the downstroke (Figure 4-34) or late in the stroke is not as damaging because the rod string is moving slower near the transition from upward to downward movement.

Tapping. Another damaging condition that can occur does so when the pump is allowed to strike either the top or bottom stops (tapping top or tapping bottom, respectively). This is sometimes done intentionally to cause the valves to jump off their seats to prevent solid accumulation or gas lock. As can be seen on the dynagraph (Figure 4-35), tapping causes sharp, heavy load changes which can overstress the entire pumping system.

Traveling Valve Checks. The dynamometer can be used to measure the action of the valves. By stopping the pumping unit near the end of the upstroke, the dynamometer can be used to determine if the traveling valve is leaking. When the valve is leaking, the weight on the polished rod load can be seen to decrease in a few minutes. A valve check is made by stopping the unit and causing the dynamometer to draw horizontal lines as shown in Figure 4-36. Since the polished rod must bear the weight of the rod string as well as the tubing fluid when the traveling valve is closed, the traveling valve check should start at a weight representing the weight of both the rod string and the tubing fluid.

Standing Valve Checks. The standing valve may be tested by stopping the pumping unit near the end of the downstroke. At this point in the stroke, the traveling valve should be open, and the polished rod should be bearing only the weight of the rod string. If the standing valve leaks, fluid begins to move

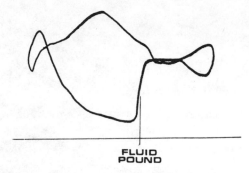

FLUID POUND

Figure 4-33. Fluid pound (the sudden opening of the traveling valve may occur near the top of the downstroke without causing severe operational problems for short times.

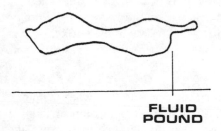

FLUID POUND

Figure 4-34. Fluid pound occurring late in the downstroke will usually cause severe equipment damage if uncorrected.

TAP

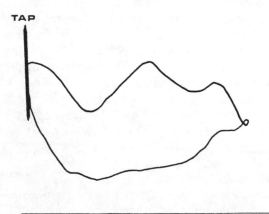

Figure 4-35. Tapping is sometimes used to keep valves clear of debris but causes severe jarring which damages equipment.

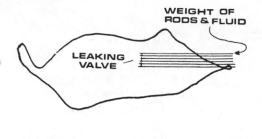

Figure 4-36. A traveling valve check determines if the traveling valve is leaking severely.

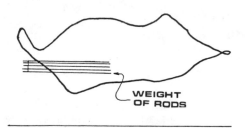

Figure 4-37. The standing valve may also be checked on a dyna-graph.

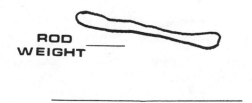

Figure 4-38. When so much gas enters a pump that the valves cannot function, the pump gas locks and does no work at all.

downward past the traveling valve. This leak will cause the traveling valve to close and the polished rod load to increase as shown in Figure 4-37.

Gas Lock. Gas lock occurs when so much gas is in the pump that on the downstroke, the gas compresses—but not to a pressure high enough to open the traveling valve. On the upstroke, the gas simply expands without allowing the standing valve to open. In other words, the pump simply does nothing, as shown in Figure 4-38.

ROD WEIGHT

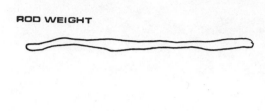

Figure 4-39. If the sucker rods break, the dynagraph shows that no work is being done.

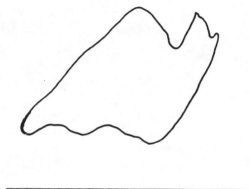

Figure 4-40. Excessive friction as a result of paraffin or scale buildup or mechanical friction causes the pumping unit to do more work than is necessary.

Parted (Broken) Rod String. When one of the sucker rods break, there is usually no doubt what has happened. However, when the rod break occurs near the end of the string of a deep well, it may not be obvious. In such cases a dynamometer can be used to determine the condition of the subsurface equipment. Figure 4-39 is a dynagraph showing a parted rod string. Note the similarity between the dynagraphs in Figures 4-38 and 4-39. Neither pump is doing any work, but the axis of a gas-locked dynagraph should be above the standing valve test line, while the axis of a parted well should be at or below the weight where the standing valve would be if it could be drawn.

Excessive Friction. When rods rub against the tubing, when the pump is dragging too much, when very viscous fluid is being pumped, or when there is not enough clearance between the rod couplings and the tubing, the polished rod load shows that more work is being done. For any pumping system, an estimate of the peak and minimum polished rod loads can be made. When the recorded loads indicate heavy loads (Figure 4-40), excessive friction may be the cause.

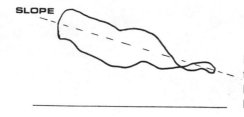

Figure 4-41. Overtravel occurs when the subsurface pump stroke length is longer than the polished rod stroke length.

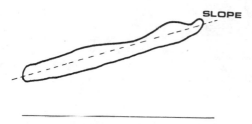

Figure 4-42. Undertravel occurs when the subsurface pump stroke length is shorter than the polished rod stroke length.

Overtravel. Overtravel describes a condition in which the rods stretch and relax, and the pump plunger actually moves a greater distance than the polished rod. When a pumping system is properly designed and operated, this is a desirable condition because it results in high efficiency. Figure 4-41 shows the characteristic shape normally associated with overtravel.

Undertravel. Just as rod stretch can cause the plunger to move a greater distance than the polished rod, under certain conditions, rod stretch can cause the plunger to move a shorter distance than the polished rod. This condition is called undertravel, and it is illustrated in Figure 4-42.

Pumping System Loading Determination

Data from dynagraphs can be used to determine loading on the pumping unit and on the rod string. By knowing the dimensions and other characteristics of the pumping system, peak polished rod loads can be converted to peak gear box torques, peak structural and beam loads, and loading of rods below the polished rod. Most pumping unit manufacturers furnish a set of data, called "torque factors," which can be used to calculate gear box loading. They also furnish structural data to allow calculation of other loads.

Figure 4-43. Acoustic fluid instruments are available to determine the level of liquid standing in the tubing-casing annulus.

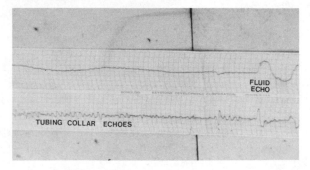

Figure 4-44. Fluid level measurement instruments chart the level of liquid in a well.

Annular Fluid Level Measurement

As was discussed in Volume 2, the level at which liquid stands in the tubing-casing annulus indicates the pressure at the bottom of a well. Since this pressure is quite important in determining the efficiency not only of a pumping system but also of the recovery system, some method is required for measuring this level. Acoustic fluid measurement systems are available for this purpose. Figure 4-43 shows one such system.

Acoustic fluid level measurements are made by causing a sharp sound in the annulus at the surface. A microphone connected to a recorder then monitors the echo of the sound wave from tubing collars, from equipment in the wellbore, and from the top of the liquid column. Figure 4-44 shows the recording made in such a measurement. By counting the number of collar echoes and knowing the length of each tubing joint, the distance from the surface to fluid can be measured to within a few feet.

Electric Power in Production Operations

Throughout the discussions of instrumentation and control, many references have been made to electric and electronic systems. In the last few years virtually all petroleum production systems have been converted to electrically operated systems. Chapter 5 briefly describes the concepts and equipment used in operating systems in this way.

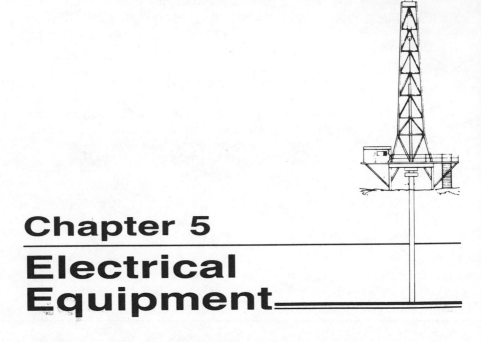

Chapter 5
Electrical
Equipment

Electrical equipment is employed on most petroleum production leases. To use electrical systems, several pieces of electrical equipment are required; to understand this equipment, some understanding of the basics of electricity is required. This chapter gives a basic introduction to electricity and is intended for the lay reader. This discussion is not, however, intended to serve as a full introduction to a study which normally takes years to complete.

Structure of the Atom

All matter is composed of atoms, the smallest entities of matter that retain the characteristics of an element. Each atom is composed of a nucleus made up of positively charged particles called *protons* and particles without a charge called *neutrons*. Orbiting the nucleus are much smaller, negatively charged particles called *electrons*. Electrons rotate rapidly about the nucleus, and the centrifugal force of this rotation would normally throw the electrons away from the nucleus, except that opposite charged bodies attract each other. The attractive force exactly matches the repulsive centrifugal force so that the electrons remain in orbit. There are as many electrons as protons in a normal atom so that the net charge difference is zero.

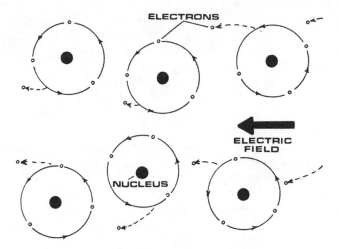

Figure 5-1. Electric current is the movement of electrons from one atom to another in the presence of an electric field.

Atoms of different elements contain different numbers of protons. For example, the element Hydrogen has one proton and one electron, while the element Oxygen has 16 protons and 16 electrons.

Atoms bond with each other to make up molecules of matter. In Volume 1 methane was said to be made up of four hydrogen atoms and one carbon atom. The methods by which atoms bond together are somewhat complex, but for the purpose of this discussion, this bonding can be simplified. In an introductory study of electricity, it is enough to describe only the bonding of a few materials.

When atoms join together to form molecules (and associated combinations of molecules), the electrons do not necessarily associate themselves with one atom. In fact, many of the electrons are shared by adjacent atoms. Some chemical combinations of atoms (such as most metals) have some electrons so closely associated with the nuclei that no normal force could disassociate them from the nuclei. When an external electric field (the same thing that causes opposite charges to attract each other) is applied, some of the loosely bound electrons can be forced to move from one atom to the next in a direction paralleling the direction of the electric field. Figure 5-1 illustrates the concept of electron movement.

A material with a large number of loosely bound electrons that can move is called a *conductor*, while a material with few electrons that can move is called an *insulator*. In this discussion conductors (such as copper, aluminum, and steel) and insulators (such as some rubbers, epoxies, and some silicon-based compounds) will be considered.

Current, Voltage, and Resistance

When an external electric field is applied to a material, some free electrons are forced to move from one atom to another. The number of electrons passing a given place in the material per unit time is defined as electric current.

An electric field is established when there is a difference in potential energy from one place to another. The potential difference is the same as the voltage difference between two points. Potential (or voltage) can be visualized as the same as the pressure that forces fluid to flow, and the current it forces can be visualized as the rate at which fluid flows under the effect of pressure.

When potential is exerted on matter, some (but not all) of its electrons will move in response. The number of electrons that will move depends on the material and on the material's temperature (as temperature increases, more electrons become loosely bound and are available for movement). This behavior represents the resistivity of any material (and its inverse, conductivity). A material is said to have low resistance (high conductivity) if a great number of electrons are available for movement, and to have high resistance (low conductivity) if few electrons are available. Using the analogies of voltage to pressure and current to flow, the conductivity of a material is analogous to the permeability of porous media.

The resistance of an object depends on its resistivity and dimensions. Since this discussion will be dealing only with wire conductors, resistance can be defined as the product of the resistivity of the material, its length, and the inverse of its cross-sectional area. That is, if the area of a wire is constant, the resistance increases with length. On the other hand, resistance of wire decreases as the wire size increases. Also, copper has lower resistivity than aluminum; aluminum has lower resistivity than steel; all three have much lower resistivity than material used for electric heating coils such as nichrome; all four have lower resistivity by many orders of magnitude than insulators such as rubbers, plastics, and glass.

The amount of current that will flow through a given media depends on the voltage exerted and the resistance of the material. In fact, Ohm's law states that the voltage across a material is the product of the current through it and the resistance of the material.

Power

Power is the manifestation of current flowing through a material as a result of voltage. Power (measured in watts) is the product of the voltage across a material and the current through it. If current is passing through material that is purely resistive, this power is given off as heat and light (heating coils

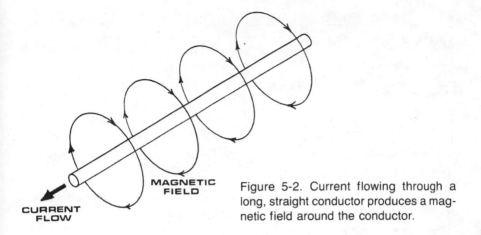

Figure 5-2. Current flowing through a long, straight conductor produces a magnetic field around the conductor.

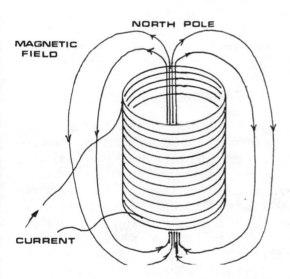

Figure 5-3. Current flowing through a coil of wire produces a magnetic field much like that from a bar magnet.

and incandescent lights, respectively). In some high-power applications the very large numbers are unwieldy, and the unit of power is measured by the kilowatt (1000 watts).

Electricity and Magnetism

It has been found that when electric current flows through a long straight conductor, a magnetic field is established around the wire as shown in Figure 5-2. If current is passed through a conductor wound in a coil, a magnetic field is established around the conductor (Figure 5-3). This magnetic field looks

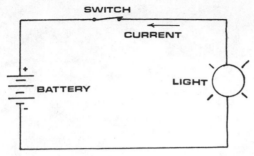

Figure 5-4. When several elements are connected in a closed loop, current flows through the circuit.

like that exerted by a bar magnet and has the same north and south poles associated with a magnet. In fact, if the coil in Figure 5-3 were wound around an iron or steel pole piece, the assembly would be an electromagnet.

Another phenomenon is that if a conductor is moved through an existing magnetic field, a current is induced in it. That is, the electrons are forced to move, but by a magnetic rather than an electric field. The same effect is noted if the conductor does not move, but the strength of the magnetic field changes.

Simple DC Circuits

Previously in this chapter, current flow has been shown to be established by a steady electric field oriented in one direction only. Such a situation is called direct current or dc because the electrons move in one direction only.

In a dc system the voltage has polarity, and electrons are said to flow from the negative to the positive side of the source. One of the most familiar sources of dc voltage is a 12-volt automobile battery. Figure 5-4 shows a simple circuit containing a battery, a light, and a switch. When the switch is closed, current flows through the switch and the conductor to the light. The voltage and current cause power to be dissipated by the light in the form of light. The resistance of the wire and switch is so low compared to that of the light that all the voltage appears across the light bulb.

If the switch is opened, there is no path for current to follow. Stated another way, the resistance of the open switch is so high that current cannot flow. When the switch is closed, a "short circuit" exists between the battery and the light; when the switch is open, an "open circuit" exists, and current cannot flow.

Series Circuits. More than one circuit element can be placed in an electric circuit. When components are arranged in series with each other (Figure 5-5), the same current flows through all elements. Now the voltage is applied across all elements, but the voltage across any element is less than the total.

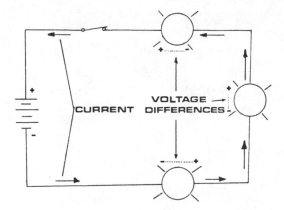

Figure 5-5. In a series circuit the same current flows through elements in the circuit causing voltages across each element.

The current through all elements is the same and is equal to the current that would be drawn from the voltage source if the circuit contained only a single component whose resistance was the sum of the resistances shown. Since the current through each element is known and resistance of each element is known, the voltage difference across each element can easily be found. The sum of the voltages across each element must equal the voltage across the entire circuit.

Parallel Circuits. When several elements are arranged parallel to each other (Figure 5-6), the voltage across each element is same as that across all the elements. The current through each element can be found if the voltage difference across it and its resistance is known. In parallel circuits the current that must be supplied by the source is the sum of the currents through the individual elements.

Alternating Current

In the previous examples the voltage was shown to be oriented in one direction only. The source for voltage can be of the type that can reverse itself. In other words, the positive and negative terminals can reverse themselves. When this happens, the electrons have no choice but to reverse their direction of flow. This situation is called alternating current or ac.

In the United States voltage is supplied from generating stations as ac, and the frequency is 60 hertz (60 reversals per second). Figure 5-7 shows a graph of voltage supplied from a generating station. Most people are familiar with 120-volt ac power supplied to their homes.

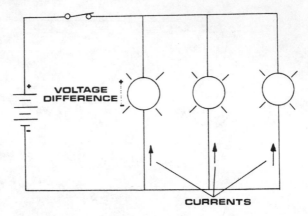

Figure 5-6. In a parallel circuit the same voltage appears across all elements, but a different current flows in each.

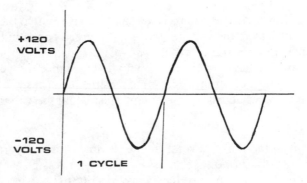

Figure 5-7. In ac circuits current and voltage repeatedly reverse their direction.

Capacitance

When voltage is impressed across two parallel conductive plates (Figure 5-8), current cannot flow across the plates, although an electric field is established across them. The area between the plates is filled with an insulator, such as air or glass, called a *dielectric*. When the voltage source is removed, it is found that the plates still have voltage across them because they have stored an electric charge. The ability of something to store an electric charge is called *capacitance*, and devices intentionally designed to

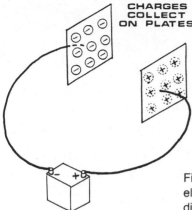

CHARGES
COLLECT
ON PLATES

Figure 5-8. A capacitor stores energy by allowing electrons and positively charged ions to collect on different plates.

store charge are called *capacitors*. Capacitors which store charge should not be confused with batteries. Batteries store chemicals which react to cause a constant voltage to be present at the terminals.

Inductance

When current is passed through a coil of wire, a magnetic field is established. When the voltage is removed, current would normally be expected to immediately cease. However, it takes a little time for the induced magnetic field to collapse. As it is doing so, the magnetic field induces current and forces current to continue to flow for a short period. The ability of anything to induce a magnetic field is called its *inductance*. The inductance of a device represents the device's ability to store energy in the form of a magnetic field like a capacitor stores energy in the form of an electric field.

Magnetic Effects in AC Circuits

Generators

Figure 5-9 shows a coil of wire inside the magnetic field of a permanent magnetic field. If the magnet is stationary and the coil is still, nothing happens. But if the coil is rotated within the field, current is induced in the wire, since the coil is moving within a magnetic field. Because the wire moves so that one side is alternately near the north pole and then near the south pole, the induced current reverses from one direction to the other and back to the original once for each complete revolution. Figure 5-10 is a graph of the current versus time of this simple machine.

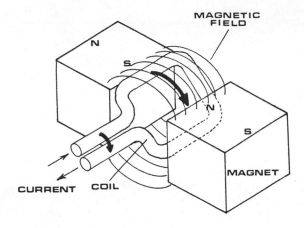

Figure 5-9. If a coil of wire is rotated in a magnetic field, a current is induced in the wire.

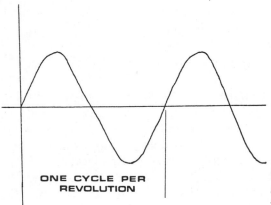

Figure 5-10. The current from a coil rotated in a magnetic field varies as a sine wave.

The device shown in Figure 5-9 is the basis of all alternators and generators used to generate electricity. The internal coil of a generator is rotated by an engine such as a gas engine or turbine, or in power plants, by a steam turbine.

Figure 5-11 shows a device with three coils within a permanent magnetic field. When these coils are rotated within the field, a current is induced in each coil. Figure 5-12 shows a graph of these currents versus time. Whereas the generator in Figure 5-9 generated a single current loop or phase, the generator in Figure 5-11 induces three-phase current. Three-phase power is the basis of most power generation in the United States.

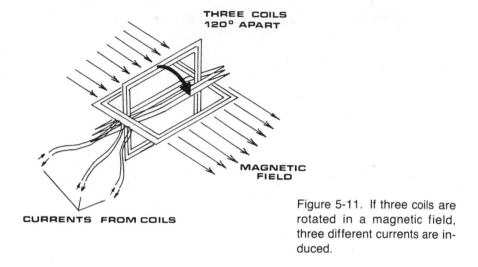

**THREE COILS
120° APART**

**MAGNETIC
FIELD**

CURRENTS FROM COILS

Figure 5-11. If three coils are rotated in a magnetic field, three different currents are induced.

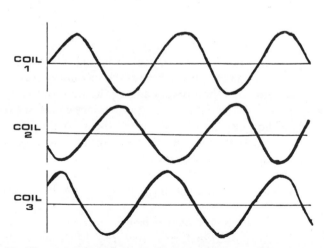

Figure 5-12. The currents induced from three coils rotated in a magnetic field are identical in waveform but displaced in time.

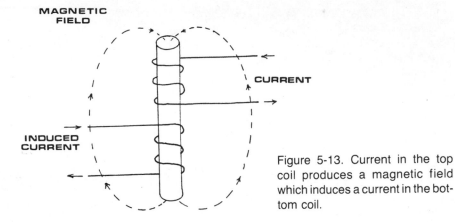

Figure 5-13. Current in the top coil produces a magnetic field which induces a current in the bottom coil.

Transformers

Figure 5-13 shows a rod wound with two coils. If dc current flows through one coil, nothing happens in the other because the induced magnetic field is steady. However, if ac current flows through one coil, the magnetic field reverses directions just as the current does. This changing magnetic field induces current in the other coil that roughly matches the original current.

When current is induced in a secondary coil, a voltage also appears across the coil's terminals. The current and voltage in the secondary coil are related to the current and voltage in the primary coil by the ratio of the number of turns in the primary coil to the number in the secondary coil. For example, if the two coils have the same number of turns, the secondary voltage and current will be about the same as the primary voltage and current. However, if there are more turns in the secondary coil, the secondary voltage will be higher than the primary voltage by the factor of the turns ratio. Also, the secondary current will be lower than the primary current by a factor of the turns ratio.

The combination of coils is called a *transformer*. The transformer in Figure 5-13 is a single-phase transformer. Figure 5-14 shows a three-phase transformer connected to a three-phase generator. The behavior of each phase in a three-phase transformer is the same as discussed for a single-phase transformer. The same effect of a three-phase transformer can be achieved by connecting three single-phase transformers together in the same way the coils are interconnected in Figure 5-14.

Most generating plants operate at a low voltage and high current. However, to transmit power over long distances, it is best to use high voltage

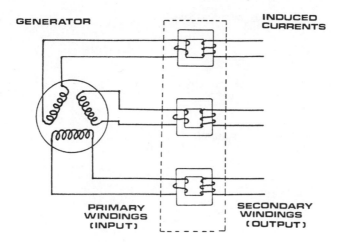

Figure 5-14. Three-phase currents may be connected to the primary windings of a three-phase transformer to induce currents in the secondary windings.

and low current because power equivalent to the product of current and the wire resistance is lost in transmission lines, and less power is lost when low current is used than when high current is used. Most power distribution systems appear as in Figure 5-15, where power is generated at low voltage, stepped up to high voltage in transformers for transmission, and stepped down to low voltage for utilization. Most homes are supplied electricity from a transformer which is fed by high-voltage distribution lines.

Three-Phase Connections

When a three-phase generator is used, its coils are usually connected together as shown in Figure 5-16. When connected in this way, the three induced currents are graphed as shown in Figure 5-17. When three-phase voltage reaches the utilization system, the three current-carrying phases must be connected to transformers in some way.

Delta Connections. Figure 5-18 shows what is called a three-phase delta connection. The three load connections, such as transformer primaries, are connected in this way.

Wye Connections. Figure 5-19 shows a three-phase wye connection. The neutral connection may be used as a current conductor for home lighting, but it is not normally connected to anything in high-voltage service.

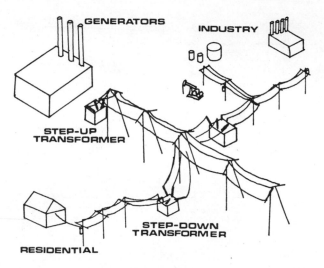

Figure 5-15. Most utility systems generate power at low-voltage. Step-up transformers distribute the power across country to step-down transformers, which reduce voltage to usable levels.

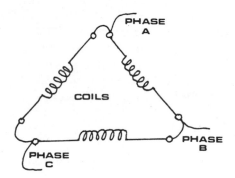

Figure 5-16. The three windings of a three-phase generator are usually connected together to avoid power distribution with six different wires.

DC Motors

Electric motors convert the energy carried by electric current into motion. A dc motor is shown in Figure 5-20. Current is passed through the stator and the rotor windings through a set of brushes or contacts arranged to assure that the current always passes through the rotor windings in the same direction, even though the supply current periodically reverses. Both sets of coils induce magnetic fields, but the windings are arranged so that the magnetic fields always oppose each other. This opposition causes motion

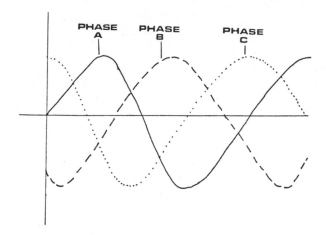

Figure 5-17. The three currents from the connected coils still appear identical except for a time shift.

just as motion would occur if the north poles of two magnets were brought together (like magnetic poles repel each other). The opposition of magnetic fields in a dc motor causes the rotor to rotate inside the stator winding and rotate an output shaft.

Induction Motors

Single-Phase Motors

The most common electric motor for production equipment is the induction motor. Figure 5-21 shows the operation of a single-phase induction motor. Current flowing in the stator winding establishes a magnetic field that induces current in the rotor winding. The current in the rotor induces a magnetic field that opposes the stator magnetic field and causes the rotor to turn. The stator windings are arranged so that there are several north and several south magnetic poles. The speed at which the rotor turns depends on the frequency of the ac current and the number of poles in the stator winding.

Induction motors are used because they are simple and durable. Note that there are no external connections from the rotor, so there are no brushes or commutator rings to wear out and require service. Single-phase induction motors are used for small loads (less the 0.75 horsepower) and are usually driven by 120-volt power.

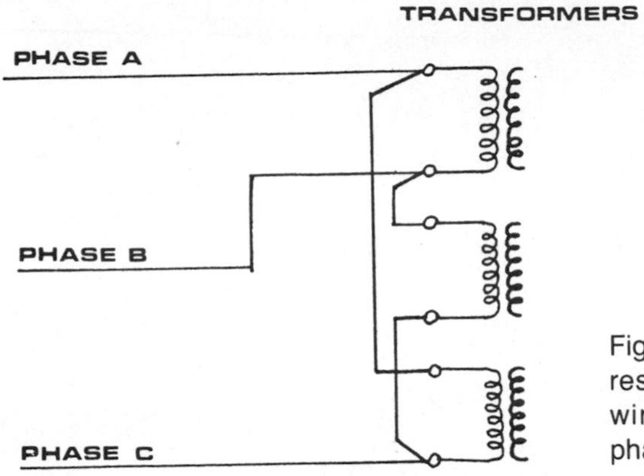

TRANSFORMERS

PHASE A

PHASE B

PHASE C

Figure 5-18. A delta connection results when two different coil wires are connected to each phase wire.

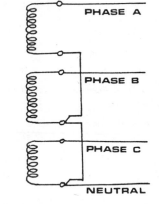

TRANSFORMER WINDINGS

PHASE A

PHASE B

PHASE C

NEUTRAL

Figure 5-19. A wye connection results when three coil wires are joined to a neutral wire and the other three coil wires are connected to phase wires.

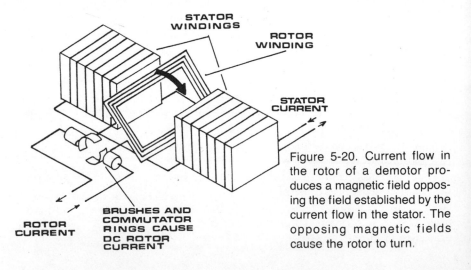

STATOR WINDINGS

ROTOR WINDING

STATOR CURRENT

ROTOR CURRENT

BRUSHES AND COMMUTATOR RINGS CAUSE DC ROTOR CURRENT

Figure 5-20. Current flow in the rotor of a demotor produces a magnetic field opposing the field established by the current flow in the stator. The opposing magnetic fields cause the rotor to turn.

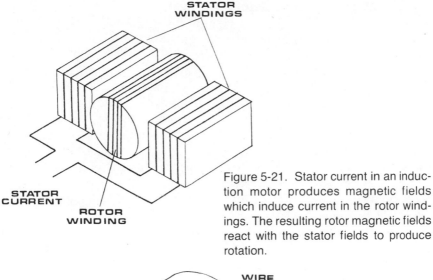

Figure 5-21. Stator current in an induction motor produces magnetic fields which induce current in the rotor windings. The resulting rotor magnetic fields react with the stator fields to produce rotation.

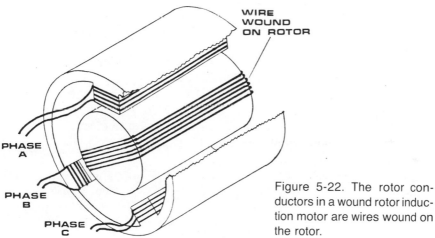

Figure 5-22. The rotor conductors in a wound rotor induction motor are wires wound on the rotor.

Wound Rotor Induction Motors

One type of three-phase induction motor is the wound rotor depicted in Figure 5-22. The stator is wound with three sets of windings—one for each phase—which are arranged to create multiple magnetic poles. The rotor is wound with an insulated conductor, usually varnished copper wire. Stator current induces magnetic fields which induce currents in the rotor windings. These rotor currents in turn induce magnetic fields opposing those induced by the stator, and the rotor turns at a speed dependent on the frequency and the number of stator poles.

Squirrel Cage Induction Motors

A squirrel cage induction motor uses the same concept as the wound rotor motor, but the rotor windings are insulated copper bars embedded in the rotor in a pattern similar to that of a squirrel cage. Figure 5-23 shows the construction of a squirrel cage induction motor. The bars are electrically connected at the ends of the rotor with insulated plates. Stator current establishes magnetic fields that induce currents in the rotor bars. These currents induce magnetic fields that turn the rotor.

Squirrel cage induction motors are by far the most common electric motors for loads greater than one horsepower. Motors of more than one horsepower are usually three-phase motors using 480-volt power. Motors of more than 250 horsepower are usually three-phase induction motors using 2400-7500-volt power. These motors are favored because they have no external rotor connections that require service and no rotor conductors which might have to be replaced periodically. The motors are hardy, durable, and dependable and can be left unattended for long periods.

Motor Classification

Induction motors have what is called a synchronous speed, which is determined by the line frequency and the number of poles. The actual operating speed is slightly less than the synchronous speed. For example, many motors have a synchronous speed of 1200 rpm and an operating speed of 1175 rpm. At loads up to the rated horsepower, the motor will operate at constant speed. When more horsepower is required of the motor than it can deliver, it slows down significantly as it overloads.

It is a characteristic of induction motors that the operating speed is less than the synchronous speed (the speed at which the magnetic poles rotate around the stator with alternating current) by a fixed fraction of the speed. This fraction is called "slip" and is constant for a motor as long as the motor operates at or below its rated horsepower. If the synchronous speed is 1200 rpm and the operating speed is 1175 rpm, the operating speed is 25 rpm (2%) less than the synchronous speed, and the motor is said to have 2% slip.

Another characteristic of electric motors is the torque they exert. Torque is a rotary force and can be determined from the horsepower and shaft speed. A 25-horsepower motor running at full load at 1175 rpm exerts a torque of about 112 foot-pounds. However, when the motor first starts, more torque is usually required to start moving it. For example, when the motor first starts, 2-3 times normal running torque may be required for a few seconds for the motor to get a pumping unit and rods moving at normal speed. Induction motors are quite good for this application because they can supply high starting torque.

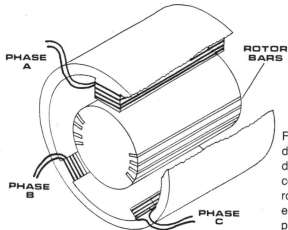

Figure 5-23. The rotor conductors in a squirrel-cage induction motor are insulated
copper bars imbedded in the
rotor and connected on the
ends with wires or connecting
plates.

The National Electric Manufacturers Association (NEMA) has set standards for electric motors so that when a particular classification is required, the same type motor will be obtained each time. The most common motors in petroleum production operations are NEMA types B, C, and D. NEMA B motors deliver a starting torque of 100-175% of full-load torque and have about 2% slip. NEMA C motors deliver a starting torque of 200-250% of full-load torque and have slip of 2-5%. NEMA D motors have starting torque of 275% or more of full-load torque and have slip of 5-10%.

NEMA C and D motors are usually preferred for cyclic loads and those requiring large starting torques such as pumping units. NEMA B motors are usually used for steady loads such as pumps. In the last few years another type of induction motor has been developed that has about the same starting torque as a NEMA D motor, but the slip in this motor is 10-20%. Thus, these motors are called ultrahigh slip motors and seem to be well suited for pumping unit applications. As the heavy loads occur, the motor slows down slightly, thereby damping some of the heavy shock loads that can cause so much damage to the system.

Electric Distribution Systems

Few petroleum production leases attempt to generate their own electricity. Generators are much too expensive to justify their expense in any but the most unusual circumstances. Most companies prefer to purchase electric power from utility companies.

When electric power is purchased for use on a lease, it must be distributed to individual wells and batteries for use. When the lease is small, a company

METERING
TRANSFORMERS

METER

Figure 5-24. Electric power from a utility company may be supplied from a substation where a step-down transformer reduces voltage and meters record power consumption.

may choose to let the utility company furnish the distribution system, but on most large leases it is more feasible for the producer to furnish his own distribution system.

Electric power must be distributed just as water used for injection is distributed. Electricity is usually distributed throughout the lease in high voltage (13,000 volts is a common distribution voltage) and then transformed down at wells or other sites.

Power may be purchased at individual sites, but when a company furnishes its own distribution system, power is purchased at a single point or just a few points. Since most petroleum leases are remote from populated areas, the utility and local distribution systems are usually overhead systems. Figure 5-24 shows the primary metering point of a major lease connection to a utility distribution system.

Overhead distribution lines are built on tall poles with cross arms and insulators to keep the bare conductors from touching the wooden or steel members. Figure 5-25 shows a common pole configuration. The wire used for overhead lines is usually aluminum strands wound around a steel reinforcement line, aluminum conductor steel reinforced (ACSR). The three phases of the electrical supply are mounted on cross arms, and an overhead shield wire that is connected to the earth is installed above the phase conductors.

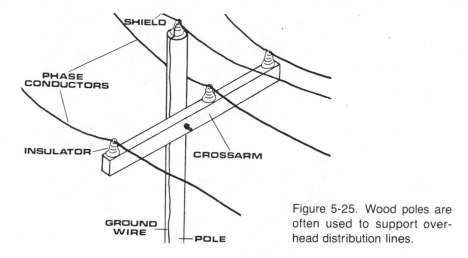

Figure 5-25. Wood poles are often used to support overhead distribution lines.

Overhead distribution systems are usually installed with switches and circuit breakers so personnel will be able to disconnect parts of the system when service is required. Circuit breakers (larger versions of the protective devices found in homes) are used to protect against short circuits when the exposed lines are damaged or overloaded. Figure 5-26 is an oil circuit breaker used to protect against overloads. Oil is used with high-voltage electrical equipment to act as an insulation and heat-dissipation medium.

Most electrical distribution systems are provided with specialized breakers called *sectionalizers*. The system is segmented, and each segment is fed by a sectionalizer (Figure 5-27). If a fault (short circuit) between phases occurs, the sectionalizer trips that segment without affecting other electrical service in the system. Reclosers are similar to sectionalizers, except when a fault occurs, the breaker trips but resets itself automatically several times in case the fault was only a momentary condition. After a predetermined number of tries, the recloser open circuits the system until it can be serviced.

Distribution lines can also be protected with simple fuses. Figure 5-28 shows a set of fused disconnect switches. If a fault occurs, the fuse elements burn out and open the line. These devices can also be used to open a section of line much like a light switch. Figure 5-29 shows the disconnect switches in the open position.

Distribution lines are often the tallest structures on a lease. When thunderstorm activity occurs in an area, the distribution lines are usually the first parts hit by lightning. Steps must be taken to protect the lines from damage and minimize power outages. Lightning always flows between clouds and the earth, and a lightning strike is seeking the path of least resistance to ground when it hits. An overhead shield conductor connected

BREAKER

Figure 5-26. Oil-filled circuit breakers provide overcurrent protection as well as a means of disconnecting high-voltage wiring.

to the ground is one way of protecting lines. The conductor provides a convenient path to the ground without allowing lightning to strike a line. Overhead distribution lines are usually equipped with lightning arrestors (Figure 5-30), devices that provide a path to ground while preventing damage to the system. Sectionalizers and reclosers are also good methods of protecting a system because these breakers open quickly in the event of a lightning surge.

Transformer Banks

Since distribution systems are normally overhead systems, transformers are also mounted on poles. A three-phase transformer arrangement is usually mounted at each point where one or more wells or utilization sites connect to the system. At each location, three single-phase transformers may be mounted in a bank as shown in Figure 5-31. These transformers are

Figure 5-27. A sectionalizer is an automatic circuit breaker that isolates the portion of a power system with a malfunction from the remainder of the system.

DISCONNECT
SWITCH

Figure 5-28. Fused disconnect switches provide overcurrent protection and a means of disconnecting electrical equipment.

Figure 5-29. When the fuse of a disconnect switch burns out, the switch opens.

LIGHTNING ARRESTOR

Figure 5-30. Lightning arrestors provide a path for lightning to reach ground without passing through electrical equipment.

oil-filled units that reduce the voltage to the level required (usually 480 volts). The transformer windings are inside the sealed tanks.

Three-phase transformers (the tanks contain all three sets of windings instead of just one set) are frequently used on new distribution systems. Figure 5-32 shows a single, pole-mounted transformer with the same capability as the three transformers in Figure 5-31. Most three-phase transformers are equipped with internal circuit breakers which protect the primary coils.

Figure 5-31. Three single-phase transformers may be connected to provide power to a three-phase system.

Figure 5-32. One three-phase transformer can provide power to a three-phase system.

Figure 5-33. A single transformer may supply power to only one motor.

Secondary Distribution Systems

Power from the secondary winding of the transformers (480 volt) must be carried from the transformer to the equipment using it. The circuits carrying this power are called *secondary distribution lines*. Sometimes a transformer supplies a single motor as shown in Figure 5-33. In this case the secondary distribution system consists of a single, three-conductor cable. At other times a single transformer may supply several motors, in which case a distribution system (depicted in Figure 5-34) is used.

Most secondary distribution systems operate at 480 volts, and the currents are much higher than those in the primary distribution system. To prevent voltage and power loss in the resistance of the secondary conductors, large wire is usually required. Also, every effort should be made to minimize the length of the secondary conductors.

Electric Motors Used With Production Equipment

The electric motors used with most petroleum production equipment are rather large, heavily constructed units. Figure 5-35 shows a large motor being used to drive a pumping unit. This motor is contained in what is called

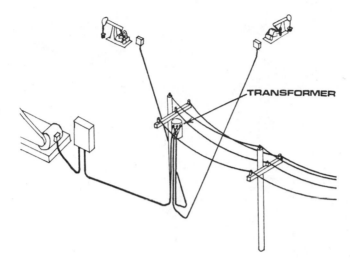

Figure 5-34. One transformer may supply power to several motors.

Figure 5-35. Electric motors are the most common method of driving pumping units.

Figure 5-36. Large electric motors are often used for heavy loads such as water injection pumps.

an open, drip-proof frame. The motor has openings that allow air circulation for cooling, but rain cannot normally enter the motor to cause moisture problems.

Figure 5-36 shows a 1000-horsepower, 4160-volt motor used to drive a large water injection pump. This motor is mounted inside a building, so it is not necessary for the motor to be protected against the entry of rain. The motor does contain internal heaters which keep water from condensing inside the motor and causing damage.

Electrical Switchgear

Electric motors cannot be operated with the simple switches used for lighting. Rather, special electrical switching circuits called *motor starters* are required for electric motors. Most motor starters are equipped with circuit breakers to protect the motors and other electrical equipment in the event of a fault. In addition, a magnetic contactor (a magnetically operated high-power switch) is used to complete the path for current to flow to the motor. Protective circuits, called *overloads,* are installed to stop the motor if it begins to draw excessive current. Figure 5-37 shows a typical 480-volt motor starter used in production applications. Since starters may be located at a distance from the motors they control, local start/stop switches are usually mounted near the motors as shown in Figure 5-38.

Slightly more elaborate starters are usually required for 4160-volt motors. First, the higher voltage is handled differently. Second, the large 4160-volt

Figure 5-37. A 480-volt electric motor is controlled and protected by a motor starter.

Figure 5-38. Local start/stop control stations may be located near the motor but far from the starter.

Figure 5-39. High-voltage electric motors require more extensive electrical equipment in the starters controlling and protecting them.

motors are much more expensive and require additional protective circuits. Figure 5-39 shows a 4160-volt starter containing all the circuits used for 480-volt starters plus circuits that detect currents leaking from the motor to ground, stator temperature detectors, and protective circuits.

Heating, Lighting, and Air Conditioning

Most production facilities require heating, lighting, and air conditioning equipment as well as some of the high-power equipment already discussed. Batteries, pump stations, and so forth must be lighted for maintenance purposes after daylight and because some security lighting is required. Most facilities are equipped with lighting transformers (480 volt to 120 volt), breaker assemblies, and light fixtures as required. Figure 5-40 shows a typical arrangement for a simple lighting scheme in a large central battery. The lighting panel is equipped with circuit breakers like those in residential panels.

The panels normally used for lighting carry 120-volt power. Actually, each of the two active conductors are 120 volts above a neutral conductor

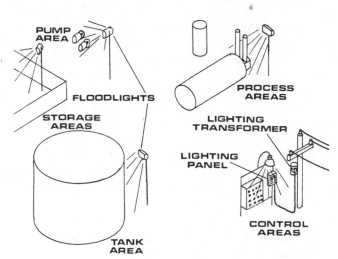

Figure 5-40. Electric lighting in a large central battery may consist of several floodlights and other outdoor light fixtures operated from a lighting panel similar to that found in most residences.

and carry 240 volts between them. Lighting fixtures and convenience outlets use 120-volt service and are connected from one active conductor to the neutral conductor. When higher voltage power is needed, such as is usually the case with electric heating units, the connection is made from one active conductor to the other to give 240-volt service.

Most lighting used in petroleum production facilities is incandescent, fluorescent (for building lighting), or sodium or mercury vapor lighting for security and general lighting. There are a wide range of lights available, and the lighting actually used must be selected for the particular application.

In addition to normal building heating, electric heating is often used in facilities which have a tendency to freeze. Water injection systems are usually built inside metal buildings which can be heated with space heaters. Of course, normal heating and air conditioning units are to make control rooms, field offices, and change rooms habitable in extreme weather. Surprisingly enough, gas heating is not often used in production facilities because these facilities are not located close to a source of natural gas that has been processed enough for interior use.

Safety

Most electrical work must be done by trained and experienced personnel only. Any attempt by untrained personnel to perform electrical maintenance work can end in injury or death.

Modern electrical equipment is enclosed in metal enclosures and conduit that are connected to the earth (grounded). This assures that even if something goes wrong inside the enclosure or conduit, a shock hazard cannot occur on the outside.

Motor starters and other electrical equipment which might require periodic service are made so that the enclosures cannot be opened without first de-energizing the equipment. This is done so that personnel not trained to perform maintenance on electrical equipment can safely replace fuses or reset breakers. Great care should be exercised to assure that the equipment grounding conductor is intact at all times. Also, water should never be allowed to stand below electrical equipment where personnel might have to work.

Electrical accidents can occur when tall equipment (well-servicing units and large trucks with gin poles) contact overhead electrical wires. Every precaution must be taken against such accidents.

Technical Improvements

Literally thousands of people are constantly involved in research and development of new practices and methods of petroleum recovery. Most of this research is aimed at improving efficiency of operations. However, there are a great number of other areas of improvement to be made. Chapter 6 describes some of the upcoming innovations in petroleum production methods.

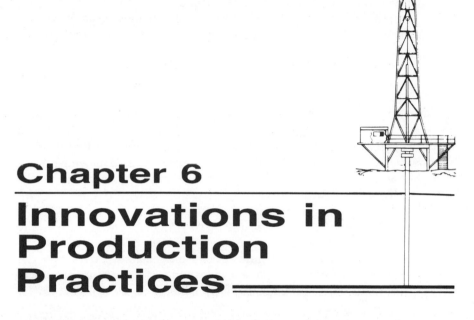

Chapter 6

Innovations in Production Practices

Discovery of additional petroleum reserves is vital to our present industrial society. Someday, alternate energy sources must takeover supplying needed energy, but until those sources can be perfected, petroleum must supply much of the energy for the entire world. Exploration efforts have been intensified in the last few years to find previously unknown reservoirs—wildcat drilling is at an all-time high. Coupled with the effort to locate new reservoirs is an equally dedicated effort to improve the methods of recovering petroleum from known productive and so-called "played-out" reservoirs.

Research has been conducted over the last 10 years to find practical methods of recovering petroleum left in the earth after known recovery methods have exhausted the reservoir. Experts believe that as much additional petroleum as has already been located can be recovered using some of these techniques. This alone doubles recoverable petroleum reserves.

In addition to trying to find new methods of recovery, producers are experimenting with new approaches to known production methods. The intent is to improve production efficiency so that additional petroleum that could not otherwise be produced can be recovered. Research into production methods does not carry the glamour of enhanced recovery

147

because the additional reserves released by such practices are not expected to approach the magnitude of those generated by enhanced recovery. However, such research could have a great impact on the consuming public, since its purpose is to reduce the cost of producing petroleum and slow the rate of cost increases.

Enhanced Recovery Injection

At this time, enhanced recovery research is proceeding rapidly. Some of the most promising methods are carbon dioxide/water miscible floods and micellar (polymer) floods. A method using a gas mixture with nitrogen is also being successfully tested. Steam injection techniques are being used in large-scale production projects to recover oil that is not much less viscous than hot asphalt.

As was described earlier, metering and control of injection wells has posed several problems. Fluids near their critical point behave quite differently from normal liquids and gases. Computer-controlled metering and control devices are often required to interpret and utilize data properly.

The injection well equipment required for some of the enhanced recovery methods is often much more complicated than that used on secondary injection wells. Often high-pressure gas injection lines must be connected to the same tubing string that is used for low-pressure water injection. Careful attention must be given to the use of isolating valves and check valves to assure that none of the pipe and fittings are subjected to more than their rated pressure. Also, care must be given to assure that gas is not injected into the water injection system and vice versa. Gas in what should be a water system causes several operational problems and can create a definite safety hazard.

It normally takes 2-5 years for production wells to "respond" to the effects of injection operations in offsetting wells. In most leases wells are usually separated by at least 1000 feet. Unique enhanced recovery testing has been conducted in that some research has been done with groups of wells only 100-feet apart. This has enabled operators to test in 1-3 years processes that might take 10-20 years to complete under normal conditions. Thus, it has been possible to test many of the enhanced recovery methods from the time of inception completely through depletion. It is possible to at least predict what will happen when these methods are used for many years.

Early testing indicates that enhanced recovery techniques have a definite effect on production practices. Under some conditions, the fluid mixtures used for recovery create emulsions that have not previously been encountered. This will require modifications in treating facility design and operation. Also, the recovery fluids enter treating facilities with produced fluids and can cause some new situations in material selection and construction techniques. Carbon dioxide is an acid gas that creates a mild acid (carbonic acid) when mixed with water, and this acid can cause corrosion problems not normally encountered. Significant volumes of carbon dioxide and nitrogen are not normally anticipated in most gas processing plants, and most are not equipped for separation of these gases. To optimize recovery activities, the injection gases must be removed from produced gas so that they may be reused and the petroleum gas may be economically processed.

It can be assumed that as enhanced recovery methods are implemented for permanent use and such systems are placed in operation, other operational considerations will arise. As dictated by the response of equipment and fluids to the effects of enhanced recovery, new designs and operating methods will have to be developed. Some may require new pieces of equipment not presently in existence.

Instrumentation Systems

By necessity, the technology of oilfield instrumentation has kept pace with the technologies developed for recovery and production of petroleum. The developments in instrumentation have required some new thinking on the part of production personnel and will continue to do so in the future.

Remote Alarm Systems

Some production operations, such as those dealing with crude processing and injection station operation, are so critical that it is important that operators be made aware of problems that occur at the time when the equipment is working unattended. Such problems include power failure, control system failure, and process failures. Many producers have installed simple electronic systems that use relays and other components to determine if proper operation is proceeding. If a problem occurs, these systems signal automatic alarm systems that notify operators of malfunctions when they occur.

Figure 6-1 shows the control panel of a simple alarm system whose only function is to turn on panel lights, a horn, and rotating beacon (not shown in figure). This system is effective if after-hours personnel can see or hear the alarms. Figure 6-2 is a more elaborate alarm system. It sends a radio message to a receiver in a nearby gas processing plant which is attended continuously. Plant personnel notify operating personnel when an alarm message is received. Figure 6-3 is an advanced form of alarm system. It dials the telephones of operating personnel until an answer is received, then plays a prerecorded message with notification of the problem.

Figure 6-1. A simple control panel may contain control circuits for several devices and warning lights, beacons, and horns to warn of malfunctions or dangerous conditions. (Courtesy of Linco-Electromatic, Inc.)

Figure 6-2. Some control systems communicate data or warn of malfunctions by sending radio messages to a central point.

Figure 6-3. More advanced control panels include control circuits and equipment to notify operating personnel of problem. (Courtesy of Linco-Electromatic, Inc.)

Electronic Instruments

There are a great number of electronic instruments in use in production operations. Many of these are complex systems necessitated by advances in production technology. There are so many examples of electronic instrument systems that it is beyond the scope of this book to attempt to describe more than just a few.

Figure 6-4 shows an electronic system used to keep track of several LACT meters. This complex system is required because three meters are connected in parallel, and they can indicate barrel counts at any time. The electronic system totals the readings from all three meters regardless of how many are running and how rapidly fluid moves through each. This particular system also monitors the meters and triggers an alarm if any of the meters fail for some reason.

Figure 6-5 shows a control system that monitors and controls the operation of several injection wells used in an enhanced recovery test operation. The operation of these wells is so critical and the data is so important that the operation is handled by this dedicated electronic system.

Figure 6-4. An electronic system can monitor the operation of several LACT units, measure liquid from several meters, and warn of problems in meters or any other equipment. (Courtesy of Linco-Electromatic, Inc.)

The system controls injection pressure and rate of injection as well as maintaining continuous meter readings reflecting pressure, injection rate, and injection volume on an instantaneous basis.

Display Systems (Annunciators)

Many production facilities have become so complex that they are operated from control boards or panels located at a central point. Operators have installed display panels (annunciators) at central points that reveal the operational status of a production system at a glance. These are panels with appropriately labeled lights that either glow steadily or sometimes blink to indicate problem areas. Figure 6-6 shows the annnciator panel for a small sulfur recovery unit. This annunciator blinks a light for the first status that causes a problem in the event of a malfunction and sounds a loud horn to call

Figure 6-5. An electronic control system monitors rates and pressures of several injection wells and operates the control valves for each well.

LAHH-511	PAH-803	BAL-305	LALL-515	LAHH-514	TAHH-716	LALL-518
ACID GAS SCRUBBER HIGH LEVEL SHUTDOWN	PROCESS AIR HIGH PRESSURE SHUTDOWN	FLAME FAILURE	WASTE HEAT RECLAIMER LOW LEVEL SHUTDOWN	WASTE HEAT RECLAIMER HIGH LEVEL SHUTDOWN	REACTOR HIGH TEMPERATURE SHUTDOWN	SULFUR CONDENSER LOW LEVEL SHUTDOWN
LAH-510	FALL-401	QA-339	LAL-512	LAH-513	TAH-705	LAL-516
ACID GAS SCRUBBER HIGH LEVEL	PROCESS AIR LOW FLOW SHUTDOWN	VALVE SAFETY INTERLOCK	WASTE HEAT RECLAIMER LOW LEVEL	WASTE HEAT RECLAIMER HIGH LEVEL	REACTOR HIGH TEMPERATURE	SULFUR CONDENSER LOW LEVEL
FALL-405	TAH-721				FLASHER	LAH-517
ACID GAS LOW FLOW SHUTDOWN	PROCESS AIR BLOWERS HIGH TEMPERATURE					SULFUR CONDENSER HIGH LEVEL

Figure 6-6. Annunciator panels not only warn of problems, but can specify where the problem occurred and in what sequence. (Courtesy of The Ortloff Corporation).

attention to the fact that a problem exists. Although this panel belongs to what is normally considered a gas processing plant, the same concepts are being used in a number of applications normally thought of as belonging to field production facilities.

Automatic Well-Testing Systems

Well testing is often a time-consuming task that has taken valuable time from production operators. Going to a battery to read meters and switch a well into test, then return at approximately the same time the next day to complete the test requires the time of people who could be more gainfully employed in analyzing problems rather than performing repetitive clerical duties. Automatic well testing is one method of more effectively using the thinking capacity of people by letting machines do routine work.

Figure 6-7 shows the control panel of a well-testing system which records the results of a test on one well and begins a test on another automatically. Unless instructed otherwise, the system tests each well for a predetermined length of time until all wells in the battery have been tested and then starts retesting each. The test header is equipped with electrically operated valves that do not require manual intervention (Figure 6-8).

"Smart" Instrument Systems

In the last few years many electronics manufacturers have begun building so-called "smart" electronics systems—circuits using microprocessors (miniature computers). These systems contain very complex electronics simplified by the use of small, integrated circuit computers which can perform a vast number of functions in mere fractions of seconds. These systems can perform simple functions from automatic well testing to operation of complex product analysis instrumentation systems. Figure 6-9 shows a system used to perform all measurements and calculations to analyze a gas stream and control the operation of a gas processing system.

Instrument System Operating Personnel

With the advent of some of the complex electronic instrumentation systems (particularly the microprocessor-based systems), it has been necessary to locate operating personnel capable of dealing with this equipment. In general, these systems require trained, highly motivated people to perform maintenance. A few years ago it would not have been possible to find such people outside military service and aerospace industries, but such highly qualified personnel are becoming more and more common in virtually all areas of the petroleum industry.

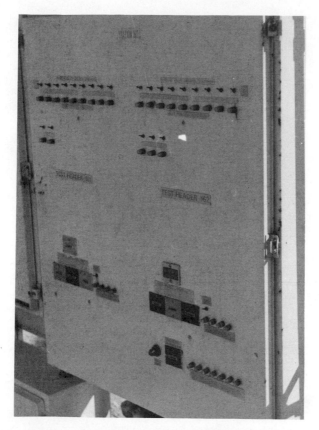

Figure 6-7. An electronic well test panel can sequentially test individual wells and keep track of fluid volumes automatically.

Supervisory Control and Data Acquisition Systems

Many production systems have become so widespread and complex that owners and operators realize that many people would be required to perform even the most routine data gathering and control functions, making the systems uneconomical. As a result, computer systems are being used with increasing frequency to collect and maintain data, perform some routine control functions, and even take nearly full control of unattended systems.

Many operators have an instinctive apprehension of computer-controlled systems (oilfield automation). Operators fear that computers will replace the people who formerly operated such facilities. This fear is, for the most part, ungrounded. Computers in even the most complex systems can

Figure 6-8. Automatic well testing requires electrically or pneumatically-operated valves that can be operated by electronic circuits.

Figure 6-9. Advanced electronic systems utilize small computer chips on printed circuit boards to control complex processes such as gas processing.

perform only rudimentary, straightforward data gathering and control functions. In fact, if such systems are properly designed and implemented, they replace only the dull, and sometimes boring, routine work. People, who have minds and can exercise judgment, are left free to analzye problems and determine solutions without the necessity of spending time reading meters and performing other manual tasks. Computers, properly used, are simply tools—perhaps expensive tools, but tools just the same.

Computer-Monitored Data Acquisition Systems

Many computer systems are installed to perform the simple task of gathering data. For example, Figure 6-10 shows the satellite battery electronics system associated with a large data acquisition system. The primary function of this system is to monitor the status of various vessels in the battery and perform automatic well testing. The computer shown in Figure 6-11 communicates with the equipment at this and about 50 other batteries and displays or prints data gathered by the well testing and monitoring equipment. On a regular basis, the machine prints reports summarizing the status of equipment. Operating personnel can observe the reports and, unless something out of the ordinary occurs, go about other duties rather than worry about this equipment. In this particular system the computer does not actually control anything but the well test headers—actual control and operation are performed by people.

Supervisory Control Systems

Some computer systems actually control activities of production systems. They turn wells on and off, open and close valves as needed, operate injection control valves, throttle the discharge of injection stations and other control functions, as well as perform all data gathering functions. Some of the first computer-controlled production systems were built in this manner, and some even performed well. Most such systems performed poorly because it was almost impossible to write the control programs (sets of instructions) needed to control equipment under all possible conditions. Another problem was that these systems were so complicated that failures often occurred in critical control functions.

Distributed Control Systems

A more realistic approach to full-scale automation has been through the use of local, electronic control systems that perform control functions without being instructed to do so by a computer system. These devices can communicate data to the computer, but control rests in the local control device which is made to operate the equipment based on particular rather

Figure 6-10. An electronics system monitors all satellite battery activities and transmits the data to a central data acquisition system.

Figure 6-11. Computer systems monitor virtually every function in some petroleum production operations and continually notify operating personnel of the status of each operation.

Figure 6-12. In distributed control systems local devices control actual operations (such as beam pumping unit control), while the devices transmit the
status of the equipment to the supervisory systems. (Courtesy of End Devices, Inc.)

than general conditions. These are called distributed control systems. Such
systems have, in general, been successful in assisting in the operation of
large, complex petroleum production systems without requiring an
inordinate amount of maintenance and intervention. Of course, most
systems are unique from all others, and some are better systems than others.

Figure 6-12 shows the local control equipment used on a producing well in
a distributed control supervisory system. The local controller performs all
data gathering and control functions and communicates this data to the
central computer system on demand. All actual control is performed by the
local system, and it makes little difference whether the computer system is
operational or not.

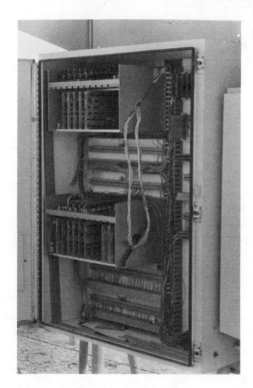

Figure 6-13. Remote telemetry units gather data from local switches, instruments, and meters and transmit the data to central telemetry equipment.

Communications in Oilfield Automation Systems

Regardless of the type of supervisory control or computer-monitoring system used, some means must be provided for communication between pieces of equipment. Data and instructions are communicated by means of a telemetry system. Each piece of equipment usually transfers data or instructions to a piece of telemetry equipment which, in turn, communicates with other telemetry equipment and the computer systems. Figure 6-13 shows a remote telemetry system which gathers data and communicates it by means of telephone cables. Figure 6-14 shows a central telemetry unit which provides communication between the computer system and the remote telemetry equipment. Some telemetry systems are so complex that it is necessary to use communication links with orbiting satellites.

Production Equipment

Production equipment has been improved drastically over the last few years to make it more efficient, less costly, and easier to operate. As was the case earlier, it is not possible to describe all the improvements made in

Figure 6-14. A central telemetry unit accepts communications from remote telemetry stations and transmits data to a central computer system.

production equipment, and the following discussion only describes a few pieces of equipment. There are many for which there simply is no room for discussion.

Although most people prefer to say that production equipment and electronic equipment are distinct systems, few mechanical systems are presently used which do not make use of electronic equipment. Some of the following equipment might be thought to fall logically into the classification of electronics when it actually belongs in the area of production equipment.

Skid-Mounted Equipment Systems

One of the greatest improvements in the construction of petroleum production facilities is the use of skid-mounted production equipment systems. A large frame of I-beams is built, and equipment such as treaters, separators, pipes, valves, pumps, tanks, and meters are permanently mounted on the frame (skid). All construction is done at a manufacturer's plant—often in a closed building so that work can proceed in any weather. When the system is completed, the entire skid (or several skids) are loaded on trucks, hauled to the facility site, set in place, connected together as needed, and made ready for operation. Skid equipment is probably the most cost-efficient equipment that can be purchased for complex systems.

Figure 6-15 shows a complete dehydrator system for a major gas production facility. All the needed equipment for the dehydrator is mounted

Figure 6-15. All equipment required in the complex process of dehydration can be mounted on skids and shipped to a site for easy installation. (Courtesy of Perry Gas Processors, Inc.)

on the skid. To have constructed such a facility in the field using discrete pieces of equipment would have taken months, while this facility was built and installed in just a few weeks.

Figure 6-16 shows a sulfur recovery unit which has been installed to serve a large gas treating plant. The equipment is installed on four skids which were hauled to the site, set in place, and connected. Again, actual construction was far faster and less expensive than it would have been had discrete equipment been used.

Figure 6-17 shows a much simpler skid-mounted system which contains three-phrase separators, tanks, pumps, and metering equipment for a satellite battery. The only additional piping that was required with this system was a well test header. Again, cost and time were saved with skid construction.

Many manufacturers offer an entire line of skid-mounted equipment. Well test systems, injection headers, complete production systems, including LACTS, and many other configurations are available on skid-mounted systems. These systems are standard "off-the-shelf" items that can be constructed and delivered in a matter of a few weeks. Special-purpose and custom-designed systems can be handled in the same way. A particular advantage of such systems is that if a skid-mounted system

Figure 6-16. A sulfur recovery unit can be completely built on skids, tested, trucked to a field location, and installed quickly. (Courtesy of The Ortloff Corporation).

is used for a time and then no longer needed, it can be transported to another location quickly and easily.

Net Oil Computers

Net oil computers are devices which use meters and analysis probes to monitor an oil/water emulsion stream continuously and record the volume of oil and water independently without ever separating them. These units are often used at satellites to test wells and monitor production instead of using treaters or three-phase separators.

Figure 6-18 shows a net oil computer using a probe which measures oil and water by measuring a property called the *dielectric constant* (a measurement of a material's ability to store energy in an electric field). This is called a *capacitance probe* and is the basis for most LACT units. Figure 6-19 is a net oil computer whose probe measures the density of an emulsion. Both of these net oil computers are capable of measuring emulsions which range from only a fractional proportion of water to those which contain more than 90% water.

Figure 6-17. A complete satellite battery can be built on a skid system to avoid building a satellite battery in the field. (Courtesy of Engelman-General, Inc.)

Figure 6-18. One type of net oil computer uses a capacitance probe to determine the proportions of oil and water.

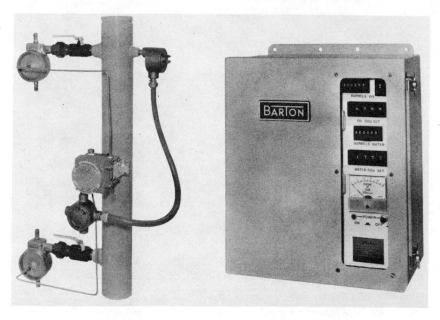

Figure 6-19. One type of net oil computer uses a ratio tube that detects density and determines the ratio or oil and water in a liquid stream. (Courtesy of ITT Barton Measurements).

Other Production Equipment

Other pieces of production equipment have been developed to improve the efficiency of production operations. One area of vital concern is the efficient use of fuel gas in fired heaters used in production systems. Vast improvements have been made in internal flow patterns and burner design to improve heat transfer. As more heat is transferred through a heater, less gas is required to maintain the same fluid temperature in a vessel. Some research and design work has been done to improve the control and reduce the amount of heat lost in exhaust gases from fired vessels. Some of these systems are so complex that they require microprocessor-based control systems, but the efficiency attained justifies the cost of such controls.

Some gas analysis equipment has been improved and made durable enough for use in production facilities. Figure 6-20 shows a gas analyzer that controls the activities of a small gas processing facility. This analyzer can be left to operate for long periods without requiring maintenance or adjustment.

Figure 6-20. Durable gas analyzers are available to control gas processing activities. (Courtesy of The Ortloff Corporation).

Evolutionary Operations (EVOP)

Large petrochemical and chemical industrial plants have for many years employed a system which is often referred to as evolutionary operations (EVOP). Such action has been applied with great success to petroleum production methods. On a regular basis, operational, supervisory, technical, clerical, and management personnel meet for a "skull session" with the intention of studying and analyzing in detail certain facets of operations. Everyone, from the youngest and least-experienced people to those who have been working for many years, is allowed to freely question every detail of an operation (leaving other operations to subsequent meetings). The intent is to ask why is an operation performed as it is and why not try something different even if there is little chance of improving efficiency. Such activities have led many producers to make changes in operating methods that drastically improved operating efficiency. The major advantage to such meetings is that they eventually cause everyone even remotely connected with production to question every facet instead of taking the attitude "we've been doing it this way for 20 years and we won't change now."

Future Innovations in Petroleum Production

Anyone attempting to look into the future of the petroleum production industry justly deserves the same reputation as some weather forecasters.

Future activities depend heavily on such changeable factors as petroleum markets, governmental regulation, economic factors, public opinion, and society demands. The only safe assumption that can be made is that even if some of the alternate energy sources are perfected in the near future, there will be a need for petroleum in abundant quantities for many years to come. Since this series was intended to introduce the field of petroleum production technology, it is fitting that some attempt be made to anticipate some near-future trends in petroleum production.

Infill Drilling

In many states regulations force the spacing of wells to be fairly wide. An average drainage area for oil wells is 40 acres (about 1.7 million square feet), and wells are required to be about 1320-feet apart. Development of secondary and enhanced recovery techniques has indicated that more efficient recovery can be attained with closer spacing. Many states allow the exception that wells can be drilled on 20-acre spacing (about 930-feet apart). Many believe that as higher production efficiencies are demanded, maximum well spacing will be reduced to 10 acres (660-feet apart). Some even go so far as to say that spacing may eventually be reduced to 1-2 acres (wells spaced 200 feet or so apart).

To decrease the spacing between wells in an existing field, wells are drilled in the center of a square made up of four existing wells. This drilling method is called infill drilling. For every reduction in spacing by a factor of two, as many wells must be drilled as previously existed. For example, a lease with 50 wells on a 40-acre spacing will have 100 wells on 20-acre spacing and 200 wells on 10-acre spacing.

There are several advantages to close well spacing. First, close spacing allows the petroleum in a reservoir to be recovered quickly. Also, close spacing often makes the formation between two wells appear more homogeneous, which makes most recovery methods work more efficiently. However, when this many wells are drilled, a great burden is placed on production facilities, and such facilities must be expanded and often redesigned to handle additional fluid production.

Increased Use of Enhanced Recovery

It is apparent that use of enhanced recovery methods must continue to increase in the next few years. The sometimes faltering steps in developing such methods will lead to efficient ways of recovering more of the original oil in place in petroleum reservoirs. Many experts expect that 60% recovery will become commonplace, and 90% recovery may be expected in the future.

Assuming that economic pressures continue, production methods must keep pace with recovery techniques. Problems with emulsions and fluid behavior that have not already been solved will be minor problems to overcome. Water disposal will become more critical because the volumes must increase with time (it is normal for water production rates to increase with time in waterflood and enhanced recovery projects). As enhanced recovery projects mature, the injection gases will be recovered at producing wells. To avoid problems in refining and to create continuing sources of these gases (such as nitrogen and carbon dioxide) injection gases must be removed from produced gas, and gas treating systems will become as vital to production operations as treaters are now.

Instrumentation and Control

Many believe that as automatic equipment is used more and more, people will be replaced by machines. This could not be further from the truth. While it is true that as production systems become more complex, sophisticated instrument and control equipment will be required, machines cannot be designed to reason out solutions to unexpected problems. People will be required to do the thinking that the machinery cannot do.

Sucker rod systems are and will continue to be the predominant lift method for many years to come. As production rates increase, many such systems will be replaced with high-volume electric submersible and hydraulic lift systems, but sucker rod pumping will continue to be the backbone of production systems. It seems now that new methods of operating and controlling pumping units appear evey day, and this development must continue. Future control methods may allow pumping units to continuously vary their speed and eliminate the need for cyclic operation, but further developments are always needed to improve lift efficiency.

Injection wells must be the heart of enhanced recovery systems. Improvements will be made in the operation and control of injection wells. Some control devices will be strictly mechanical, but most will be electronic and/or pneumatic in nature.

It is already becoming difficult to draw a line between mechanical equipment and the control systems that operate it. Mechanical equipment such as treaters, separators, pumps, compressors, and so forth must continue to improve. It is likely that more automatic control and instrumentation will also be required to maintain high operational efficiency.

Petroleum Production Technology

Several years ago, during the first critical gasoline shortage, a lawmaker was asked his solution, and he said that the petroleum producers should simply go drill more gasoline wells. Whether or not his comment was in jest, the comment illustrates the point that few laymen understand the intricacies of petroleum production, let alone the petroleum industry in general. It is hoped that this series has adequately introduced petroleum production to those not involved in the industry, and perhaps has raised a few points to those who are. Throughout this series, an emphasis has been placed on optimization and efficiency. This is not a foggy technical concept, but an absolute necessity for petroleum production technology. Even if regulations do not require higher efficiency in recovering and processing petroleum, market demands, economic conditions, and society pressure will.

Index